Energy Centered Management:
A Guide to Reducing
Energy Consumption and Cost

Energy Centered Management: A Guide to Reducing Energy Consumption and Cost

Marvin T. Howell

Routledge
Taylor & Francis Group

LONDON AND NEW YORK

Published 2020 by River Publishers

River Publishers

Alsbjergvej 10, 9260 Gistrup, Denmark

www.riverpublishers.com

Distributed exclusively by Routledge

4 Park Square, Milton Park, Abingdon, Oxon OX14 4RN

605 Third Avenue, New York, NY 10017, USA

First issued in paperback 2023

Library of Congress Cataloging-in-Publication Data

Howell, Marvin T., 1936-
    Energy centered management : a guide to reducing energy consumption and cost / Marvin T. Howell.
        pages cm
    Includes bibliographical references and index.
    ISBN 0-88173-746-1 (alk. paper) -- ISBN 978-8-7702-2324-9 (electronic) -- ISBN 978-1-4987-3692-3 (Taylor & Francis distribution : alk. paper) 1. Energy conservation. I. Title.

    TJ163.3.H685 2015
    658.2′6--dc23

                                                                    2015006975

Energy Centered Management : A Guide to Reducing Energy Consumption and Cost / Marvin T. Howell.

First published by Fairmont Press in 2015.

Routledge is an imprint of the Taylor & Francis Group, an informa business

Publisher's Note
The publisher has gone to great lengths to ensure the quality of this reprint but points out that some imperfections in the original copies may be apparent.

ISBN 13: 978-87-7022-932-6 (pbk)
ISBN 13: 978-1-4987-3692-3 (hbk)
ISBN 13: 978-8-7702-2324-9 (online)
ISBN 13: 978-1-0031-5193-7 (ebook master)

While every effort is made to provide dependable information, the publisher, authors, and editors cannot be held responsible for any errors or omissions.

The views expressed herein do not necessarily reflect those of the publisher.

Dedication

This book is dedicated to the United States Air Force, where the author had the privilege of serving 20 years in the civil engineering career field and retiring as a Lt. Colonel, and to the Association of Energy Engineers who the author has been privileged to participate with as a trainer in their online energy training program. This book is also dedicated to any organization worldwide that has selected as a goal to reduce energy use at their facility or facilities. Finally, this book is dedicated to my wife, Jackie Howell, and my oldest grandson, Christopher Cline, who helped me with the figures and charts.

Table of Contents

Preface . . . . *xi*

Chapter 1
Energy Centered Management System 1
    Energy Centered Management System (ECMS) 1

Chapter 2
Energy Centered Planning and Development (ECP&D) 41
    Purpose. 41
    Energy Policy 47
    Energy Plan 50
    Appointing a Management Representative 51
    Establishing an Energy Team 51
    Establishing Energy Performance Indicators and a Baseline . . 54
    Corporate Procedures 60
    Energy Training 60

Chapter 3
Energy Centered Waste (ECW) 61
    Identifying Energy Waste 61
    Energy Walkthrough(s) 63
    Performing an Energy Audit 65
    Energy Awareness Training 68

Chapter 4
Energy Centered Objectives (ECO) 71
    Some Words that Cause Confusion 71
    Where Do Objectives Come From?. 72
    Energy Objectives and Targets and
        Making Targets or Objectives SMART 85
    Don't Chase Too Many Rabbits 87

Chapter 5
Energy Centered Projects (ECP) 95
    Introduction 95

Possible Projects 96
Energy Services Performance Contracts (ESPC) 97

Chapter 6
Energy Centered Maintenance (ECM). 99
ECM Elements 99
Why ECM? 100
Objective of ECM 100
O&M Savings 100
ECM Aim, Components, and Design 101
The ECM Phases 104
Using Reliability Theory 120
Applying The ECM Process—Getting Started—
   The Steps to Take 122

Chapter 7
Energy Reduction Deployment and ISO 50001
Energy Management System 139
Introduction 139
Energy Reduction Deployment Outline and
   What is Needed to Conform with ISO 50001 EnMS 139
ISO 50001 EnMS Would Require Additions 141
Superior Energy Performance 143

Chapter 8
Self Inspection and Internal Audits 149
Introduction 149
Gap Analysis 150
Excellence Factors Evaluation 153
Self Inspection/Internal Audits Checklist 164

Chapter 9
Creating an Energy Reduction Culture and Emphasizing
Energy Conservation 187
Developing an Energy Reduction Culture 187
Developing and Implementing an
   Energy Conservation Program 193
Guiding Principles 193
Energy Conservation by Areas of Impact 194

Chapter 10

IT Power Management 199
What is Power Management? 199
Purchase Energy Star Computers,
    Monitors, and Other Electronics. 199
Enable Energy Star Sleep Function 201
Minimize Screen Savers. 201
Reduce Time of Idle 202
Ensure Energy Star Use and
    Turning Equipment Off When Not in Use 202

Chapter 11

Reducing Office Paper Use 205
Objective 205
Implementing the Major Objective:
    Reduce Office Paper Use and Save Energy 206

Chapter 12

**Energy Centered Maintenance (ECM) & Energy
Centered Projects (ECP) in Data Centers** 213
Establish an Energy Reduction Team 213
Key Definitions 213
Energy Team Activities 214
Identifying Energy Waste. 215

Chapter 13

Building Your Energy Reduction Plan 221
Quick Method 221
The Process. 223
Traditional Method 224
Determining Contributions Needed to Achieve Goal 228
Savings or Cost Avoidance Verification 232
Creating the Organizational Culture—An Assessment 235

Chapter 14

**Drivers of Energy Reductions and Continuous
Improvement & Verifying Results**. 241
What is a Driver? 241
The Drivers of ECMS 241

Continuous Improvement 242
A Barrier or Non-driver Which Must Be Considered 244

ECMS Glossary 247
Definitions 247

Bibliography 255
Books/Articles 255
Websites 256

Index 259

Preface

This book will provide any organization or company a guide to use in planning, developing, and implementing an energy reduction and management program. It is specially designed to achieve energy reduction deployment from top management to all employees and onsite contractors. Energy reduction deployment (ERD) can be implemented by itself and render savings in the 10-20% range. However, for a reduction goal of more than 20%, an organization will probably need to identify major energy projects. To deploy energy reduction from the top of the organization to the lowest employee, a new system, energy centered management system (ECMS) was developed. The ECMS can be implemented and be in congruence with the ISO 50001 Energy Management System (EnMS) phases and elements and meet the additional requirements of SEP (special energy performance). The missing emphasis and functions of ISO 50001 EnMS provided by this book are:

1. A method to deploy the energy policy throughout the organization to include the planning, developing and organizing needed.

2. A comprehensive approach to identifying energy waste in the organization.

3. A component showing the development of objectives and targets by an energy team and then the development and implementation of action plans when implemented that will achieve the objective and target.

4. A component explaining the identification and development of energy projects that will reduce energy significantly and with a short payback period.

5. A method to prevent and/or correct problems caused by poor or inadequate maintenance and/or energy waste conditions.

The hunt for energy waste can consist of management/employee brainstorming, functional teams addressing what energy waste exists in their area, walkthrough by experienced energy personnel of their facilities, energy audits conducted by professional(s), and research of products and technology. Once the waste is identified (energy centered waste), then it is fixed by corrective or preventative maintenance (energy centered maintenance), by an objective or target (energy centered objectives), or an engineering project (energy centered projects). If waste is found with equipment going to a state where it is still functioning but using excessive energy, then the organization employs energy centered maintenance (ECM). If an organization has T-13 lights and researches lighting options, it will find that either T-8s or T-5s are available and reduce energy waste significantly if they replace the old fluorescent lights. Then installation of T5s would be a good energy project. The new lights will have only one electronic ballast replacing two obsolete ones per lighting station. The waste was found through research and the fix is via energy centered projects (ECP). If during the walkthrough lights are left on in restrooms, hallways, break rooms and mechanical rooms, and no one is present, this energy waste is best handled through the energy team establishing an objective and target and developing an action plan (ECO) that includes first determining the number of occupancy sensors needed, whether they should all be done at once or phased in, how much energy they will save, and the payback period for alleviating their cost.

This concept is shown in the following figure.

ECMS (Energy Centered Management System)

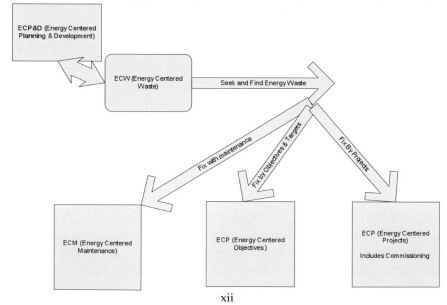

Presently, there are three energy management systems available to be used separately or use superior energy performance after ISO 50001 EnMS has been implemented. The Energy Star EMS is a simple approach that beginners will find useful and can build on to a higher EMS implementation later. It does provide a portfolio management system that organizations can use to record their energy use and building information, and receive a score, which is a benchmark of how their facility stacks up with other similar ones. ISO 50001 Energy Management System is an excellent system that is benchmarked from ISO 14001 Environmental Management System which has proven very useful in improving environmental performance. SEP starts with ISO 50001 EnMS implementation, adds a few minor requirements to the 23 elements, and then requires personnel that work with organizations to be trained as practitioners by either AEE or Ga. Tech. ANSI controls what organizations can certify the organizations as to meeting the SEP criteria. The SEP criteria are a simple approach and several manufacturing companies have reduced energy significantly by applying this approach.

One of the innovative requirements is to verify results by ensuring the results compared to an established baseline are real and the weather changes and production levels are adjusted to determine that real results have been achieved. Presently, only manufacturing and water works companies are eligible to participate. DOE is projecting that others may be eligible in about two years.

The three systems are helpful, but in reducing energy at an organization, there are several tools, techniques, and items that can enable an organization to improve its implementation of an EnMS and increase its energy performance. These are not covered or even mentioned in the above approaches to improving an organization's energy performance. This energy centered management system (ECMS), provides helpful tools and guidance. ECMS supplements well the three approaches explained above and will not duplicate any of the three systems' requirements while meeting or satisfying them.

ECMS is the only EnMS that includes:

1. Developing an energy waste list using five different methods to help build the list.

2. Measuring organization and machine energy efficiency.

3. Have four components available to minimize or eliminate the energy waste.

4. Provides an array of energy metrics that enables any organization to select meaningful energy metrics for their EnMS.

5. Includes an energy centered maintenance component that was designed using energy management and reliability management principles and techniques.

6. Provides a method and process for estimating how much a countermeasure will contribute towards meeting the organization's goal.

7. Identifies the normal existing low hanging fruit plus encourages innovation in developing energy waste reduction.

8. Includes the principals and tools to help an energy team achieve high performance status.

9. Spells out top management and leadership roles and responsibilities and provides an assessment tool to see how they are doing in your organization.

10. Includes methodology and actions essential to move your organizational culture to one of energy awareness.

11. Uses tools/techniques to enable implementation and achieve excellent energy performance. Some of the tools/techniques are critical success factors, excellent factors, energy reduction checklist, leadership assessment, organization cultural assessment, energy audit, and a self-inspection checklist that can also be used to conduct internal audits.

12. Includes the problem solving process and tools to prevent energy performance or energy management system failure or chasing too many rabbits.

Simple to Understand and Easy to Apply

For organizations, that do not at this time want to become certified but want to reduce energy consumption and cost efficiently and effectively, ECMS (energy centered management system) is a viable alternative and an easy-to-use energy management system. Later, if certification becomes an objective, then ECMS documentation and actions can be changed to comply with ISO 50001 Energy Management Systems elements, and certification can be pursued.

The book is designed to present the ECMS in Chapters 1-6. Chapter 7 covers what additions are necessary to have ECMS conform to ISO 50001 Energy Management System (EnMS). Chapter 8 provides the checklist and information on how to perform an internal audit or self inspection. Chapter 9 discusses how to create an energy awareness organizational culture. In Chapter 10, IT Power Management, shows how to implement the benefits and probable results. Chapter 11, Reducing Office Paperwork, shows a simple ECO example of how energy can be reduced and cost savings achieved. Chapter 12 includes ECM in data centers and demonstrates the potential for all four of the five subsystems (ECP&D, ECW, ECP, ECO & ECM) to be used with ECP being the most probable fix to reduce energy and achieve the vast energy savings possible in our data centers. Chapter 13 demonstrates how to build an energy plan. Chapter 14, Drivers of Energy Reduction and Continuous Improvement, shows 10 drivers that can help any organization achieve energy reductions and maintain the momentum through implementation and to continuously improve their processes through the energy management journey.

Chapter 1

Energy Centered Management System

ENERGY CENTERED MANAGEMENT SYSTEM (ECMS)

In the United States, around $500 billion a year is spent on energy. In the world, industry consumes 51% of total energy produced. Energy costs represents up to 30% of corporate operating expenses. The U.S. Green Buildings Council estimates that commercial office buildings use, on the average, 20 percent more energy than they need to do. This is an astounding dollar loss to industry due primarily to the fact that management does not know where the waste is occurring and what to do to eliminate or reduce this loss. Reducing energy costs is a major opportunity for most organizations and/or companies today. The 20% more energy than needed is the energy waste that exists. The purpose of this book is to give organizations' management a roadmap to reduce energy in an efficient and effective manner. ISO 50001 Energy Management System (EnMS) and Department of Energy's Superior Energy Performance (SEP) give organizations a helpful guide for managing energy use, consumption and cost. However, they do not provide a strategic deployment strategy, method or system to do so. The energy reduction deployment model/process called energy centered management system is designed to deploy the energy policy from top management to every employee and contractor located on site. In designing the energy centered management system, human factors (or the stakeholders' voice) were considered. ECMS borrowed from or incorporated concepts, methods, and principles from policy deployment, organizational behavior and change, metrics development process, critical success factors, reliability centered management and ISO 50001 Energy Management System. It does not require any duplication of work to implement ISO 50001 Energy Management System. In fact, it enhances ISO 50001 Energy Management System implementation.

In implementing any strategic initiative or objective, the human factors should be considered and addressed to ensure success. The internal players are top management, middle management, and employees and contractors assigned to the facility. Each has a stake in this effort and their thinking and motivation are different on energy reduction.

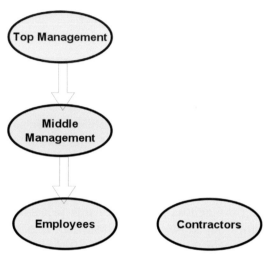

Figure 1-1. Human Thinking and Motivation—The Major Participants

The thinking and motivation are different for each participant; therefore, addressing their actions or considerations will differ. First, let us look at top management's thinking.

The questions on the minds of top management need addressing to get energy reduction kicked off and on a path to becoming a reality. A consultant, internal or external, would be helpful at the beginning,

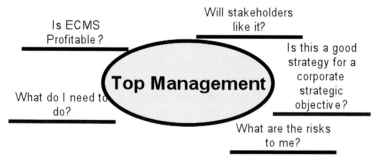

Figure 1-2. Human Thinking and Motivation—Top Management

although a diplomatic and competent energy manager should be able to accomplish this. He or she could gather energy bills for a few years, plot the data of energy use and cost, and then show what a 5%, 10%, 15% and 20% reduction would save in energy cost. The stakeholders are the board of directors (if the organization has one), their headquarters, the public, and their management and employees. Each of these would have different interests such as how much greenhouse gasses will be reduced, how disruptive will the program be to the organization, are adequate resources available, and what will be required of me if the effort is launched. The major potential cost savings will satisfy most of the stakeholders and gain their initial support.

The consultant or energy manager can explain that other organizations have done this as a strategic objective and have been very successful. The risks are minimal if top management supports the effort and doesn't just give lip service to it. The program will be successful, and they will get the credit for initiating it and supporting it to fruition.

What does top management need to do? They need to:

1. Appoint a management representative who will become their energy champion. He or she will run the day-to-day efforts and keep top management informed of progress and results. The energy champion should be a member of management and committed to reducing energy consumption and cost. One of the first duties of the energy champion will be to establish an energy team comprised of people from all major functional areas, thus making it a cross-functional team. The energy champion and energy team should develop a draft of the corporate objective or goal with a target for achieving a compelling energy policy for the organization. The energy champion will need to meet with top management and get their approval of both.

2. Top management should communicate the corporate objective/goal, the energy policy and why their achievements are important to the organization. They should encourage all the organization's people to support the energy reduction initiative in any way they can. Top management should take advantage at any meeting to express their support for the energy reduction program.

Next, middle management does have a role in this important effort. See Figure 1-3 for their thinking and possible motivation.

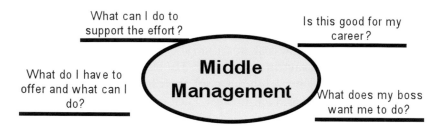

Figure 1-3. Human Thinking and Motivation—Middle Management

Middle management wants to know what they can do. It is going to be up to the energy champion to let them know. First, some of them can serve on the energy team and the walkthrough of the facility or facilities. Once the energy conservation training is given to all the personnel in the organization, they can lead and encourage conservation efforts. They can assist in implementing IT power management and reducing paper use by duplexing and increasing electronics files. These initiatives will be discussed further in later chapters. It is good for their careers if they are recognized for their involvement and contributions. Their boss wants them to be involved and continue to do their full-time job well. They do have a lot to offer. The energy champion and energy team will tap them occasionally to help accomplish an objective or target some other essential tasks.

What about the employees? Figure 1-4 shows their original thinking.

The energy champion and energy team should develop energy awareness training and give it to all management, employees and contractors. An easy way to accomplish this is to develop a PowerPoint pre-

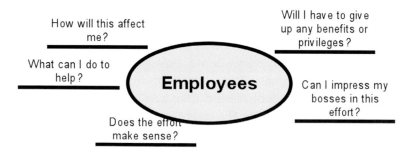

Figure 1-4. Human Thinking and Motivation—Employees

sentation that includes some pictures at the facility showing significant energy users and waste. The training could include an energy conservation program that all should participate in.

The benefits and privileges issue can be a touchy one that if not handled properly can lead to a morale problem. Once employees are allowed to bring electronics or appliances in their areas such as radios, stereos, microwaves, fans, heaters, and small refrigerators, they feel entitled to them. Programs to get them to turn off and unplug them at night, on holidays and weekends or even to remove some from the work place can be met with resistance. The energy champion must communicate prior to any actions why these must be addressed. Another example, that everyone takes for granted until a change is made is the thermostat settings at the facility. The recommended settings are 78 degrees for the summer and 68 degrees for the winter. These settings may require dressing changes such as needing a sweater or jacket in the winter and summer clothing when it is hot. The employees have to be conditioned for any thermostat changes. They have become used to the degrees being around 72 year round. They will immediately feel any change, and it does affect their productivity. The energy savings in just one degree change is significant, so this action should always be considered by the energy team with the support of top management and middle management. If a decision is made to do this, then management must create the convictions to accept this type of action to save money for the organization that can be used to fund other vital mission areas and needs.

This type of effort does enable employees to get involved in an environmental and energy initiative that is good for the organization and the community. It is an opportunity to impress supervisors with their attitude, involvement and contributions. It is important to get the energy awareness and conservation training to them as fast as possible.

Figure 1-5 shows the contractors' initial thinking on this new program.

Figure 1-5. Human Thinking and Motivation—Contractors

The contractors have a contract that outlines what they can work on and when they must finish deliverables. Naturally, when they hear this they want to know what is expected of them, and will it change their contract. Also, they want to help. The energy awareness and conservation training should also be given to all on-site contractors.

There are two important external stakeholders for companies. The board of directors (if there is a board) and the organizations' customers. The board of directors will be pleased and supportive since the initiative is being accomplished with the desire to reduce energy costs. The customers will be supportive since most support the environment, and saving energy and lowering their costs could result in savings being passed on to them when they buy the company or organization's services or products.

The above voices when heard and considered will get all participants involved and focused on reducing energy. These views were factored into the energy centered management system's design and should be considered in implementing energy reduction actions.

Top management starts the corporate initiative when they decide on the need to reduce energy consumption and energy cost. It must be written in a corporate goal or objective format that shows how much, the energy type (electricity, natural gas, steam, etc.) and the time expected to achieve the goal. Next, there has to be a system, processes, or a model to use to make it happen. Just deciding that it is needed will not get it done. ISO 50001 Energy Management System is an excellent standard mirrored after the successful ISO 14001 Environmental Management System. SEP adds a few requirements to ISO 50001 EnMS with the largest being the measurement and verification of energy performance results. However, they do not provide a clear deployment process or system other than the continuous improvement cycle of plan-do-check-act. This cycle is very important and used in so many ways for improvement, but it is not descriptive enough to serve by itself as an energy reduction deployment method without the organization adding a lot of specifics.

What should an energy reduction deployment model look like? There are some parts we know that exist in policy deployment and/or strategic planning: leadership, top management support, measurement, strategies, and employee involvement to name a few. Of course, they will be included in any model. Energy waste needs to be identified, and then there must be avenues to fix or eliminate or minimize the energy waste. There should be ways to check progress and verify results. To be complete, the model should contain an avenue through review to continuously improve the energy reduction deployment. Like policy deployment (en:Wikipedia.org/wiki/Hoshin Kanri) the model should 1. Focus on a shared goal, 2. Communicate to all participants, 3. Be able to measure progress and results, and 4. Hold accountable all participants for achieving their part of plans. The model is shown in Figure 1-6.

Each of the model items and/or components will be discussed separately. The main components are shown in Figure 1-7 ECMS Components.

Each of these components will be covered in a separate chapter. They represent in simple terms that certain things have to be done to reduce energy. The EP&D component is to obtain management commitment and involvement; set goal(s); develop and communicate an energy

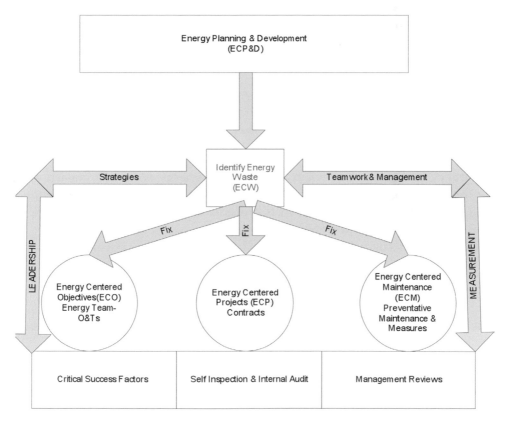

Figure 1-6. Energy Centered Management System

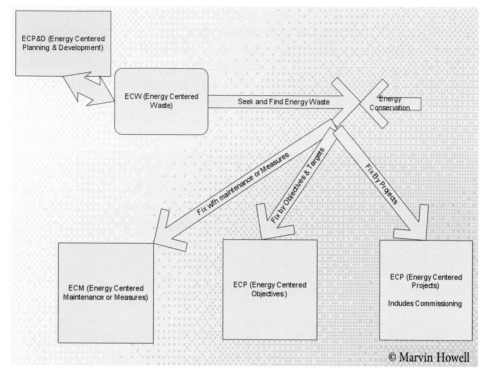

Figure 1-7. Components of ECMS (Energy Centered Management System)

policy; do an energy review and establish energy metrics to measure progress and results; established a baseline so results can be identified; develop an energy plan; and write procedures to cover the requirements for the processes and requirements.

In this chapter, the essential characteristics will be covered. These are necessary to ensure favorable results by providing the management essentials. They are leadership, strategies, measurement and monitoring, and teamwork and management functions such as project management, budgeting, and project approval.

The foundation includes critical success factors that help an energy team improve its performance: internal audits, self inspections, and management reviews. All three enable an organization to continuously improve its energy management performance. The management essentials will be outlined in this chapter.

First, leadership will be explained as to why it is included.

Leadership

Seldom will any objectives deployment, strategic plan or improvement action succeed without leadership.

- Top management including the energy champion must provide leadership throughout the energy reduction deployment, starting with development of the corporate goal and energy policy and communicating it and their support to all of the organization's personnel.

- They must communicate clearly to motivate people, to enable focus, and for enthusiasm in support of the energy reduction initiative to prevail.

Their confidence, optimism, inspiration and empathy are needed to keep everyone working together to achieve the corporate goal and energy policy. Leadership can be thought of as the most important organizational glue to keep the organization focused. Their importance could be described as the GLUE (great leadership in using energy).

A leader is someone whose direction and vision other people are willing to follow. Therefore, GLUE leadership is defined as:

"Influencing others to follow a given direction and vision."

"Leaders provide the glue to make a vision and/or policy to have followers who stick together to achieve the vision or policy by following energy centered management system guidance."

GLUE leader's top 10 skills must be:

1. Creating a long-term compelling energy policy.
2. Communicating the goal or objective to other top managers, middle managers, supervisors, teams, individuals and on-site contractors.
3. Creating strategies and an energy plan.
4. Inspiring, motivating, and influencing others to act to meet goals and objectives.
5. Reviewing performance, progress and results and communicating direction.
6. Facilitating change and building consensus.
7. Encouraging continuous improvement of energy reduction processes.
8. Developing teams and individual talents.

9.   Driving achievement and performance.

10. Reviewing and celebrating success by recognizing teams' and individuals' efforts and rewarding exceptional performance.

Leadership Assessment

Leadership is a necessity for any improvement initiative to be successful. This includes energy reduction. Leadership must provide the guidance, motivation, emphasis and support for energy reduction to be successful. How effective are your leaders at improving our learned habits and focus on reducing energy waste? Please fill out the Leadership Assessment Survey to determine if they are ready or if coaching and training are going to be needed.

Not Good   **Very Good**
  1   *2*   *3*   *4*   *5*   *6*   *7*

1.   Our top management has developed a clear, compelling energy policy and has communicated it well. Score____

2.   Our top management is good at long-range planning. Score____

3.   Our top management inspires managers, employees and contractors to participate in energy conservation efforts and in identifying energy waste. Score____

4.   Our top management expects excellent results from our energy deployment efforts. Score____

5.   Our leader's or leaders' actions reflect our core values. Score____

6.   Our leader or leaders know what it takes to be successful. Score____

7.   Our leaders respect people's feelings. Score____

8.   Our leader or leaders are optimistic and can overcome barriers and road blocks to get our reduction effort back on track. Score___

9.   Our top management challenges all to learn new skills and knowledge. Score___

10. Our top management appreciates superior performance and rewards and recognizes these efforts. Score____

11. Our top management is good at delegating work and responsibilities. Score___

12. Our leaders are good at communicating, since they speak clearly and explain things well. Score____

13. Our top management knows what goes on in our organization. Score____

14. Our top management thrives on change. Score___

15. Our top management keeps up with the progress the organization is making toward achieving the energy reduction goal. Score____

16. It is obvious that our top management is committed to reducing energy consumption and costs. Score____
17. Our culture is fast becoming one of reducing energy consumption and costs. Score___
18. The energy champion is on top of all problems and continuously emphasizes to all how they can help in reducing energy waste. Score___
19. Our leaders readily listen to new ideas, suggestions and recommendations for improving the energy reduction initiative. Score___
20. Top management emphasizes how all can contribute to reducing energy consumption. Score____
21. Top management is committed and involved in energy reduction. Score____

- 21 questions @ 7 points each gives 147 points maximum
- 21 questions @ 1 point each gives 21 points minimum

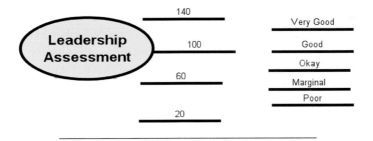

If the results are above 75, then the leadership should be adequate to help drive the energy reduction. If below 75, then coaching by a facilitator and developing training for them may be good to have in the energy plan or least an objective and target for the energy champion and energy team.

Strategies

Strategy or strategies are very important in any improvement effort.

Definitions & Purpose:

— A plan of action or policy designed to achieve a major or overall goal.
— A high level plan to achieve one or more goals under conditions of uncertainty.
— A strategy determines how a goal will be achieved by the means (resources).

Method:
— 1. Set goal.
— 2. Determine actions to meet the goal.
— 3. Monitor the resources expended to achieve the goal.

A strategy is a plan, a method or series of maneuvers for obtaining a specific goal or result. For energy reduction, four strategies come to mind almost immediately. To use the strategies to achieve the energy reduction goal, the following actions must be taken.

Strategy 1—Use the energy reduction deployment model.

Strategy 2—Develop energy awareness training and give to all personnel. Develop energy conservation training and present to all personnel. Leadership must communicate the need to identify energy waste before we can fix it. "We need all of you to help identify the energy waste in our organization. You do that, and we will work on eliminating or minimizing it."

Strategy 3—Get everyone involved. The training in strategy 2 will help. Communicate at every opportunity. In developing the energy conservation program, conduct employee "brainstorming sessions" to identify how they can reduce energy consumption, use and cost. Roll their ideas into the program and then train them on the program. Have an energy manager, the energy champion, the energy team leader, or energy facilitator cover the energy awareness training prior to the brainstorming session. Set up some functional teams or natural working groups to identify waste in their functional areas. Going forward, let everyone know how they can help and keep them informed of progress and results achieved along the path to meeting the goal.

Strategy 4—Contact the utility and see if they provide funding support on reducing energy use. Learn the program and develop reduction projects and present to the utility. If government-related, check out the large engineering and construction companies that fund projects (energy services performance contracts), construct or implement them, and they will get paid for their efforts through the savings the organization incurs from the project(s). This program has been very successful for government facilities, including all of the Department of Defense.

Strategies Selection

Normally four to six strategies will suffice. A six-strategy approach is shown in Figure 1-9. Strategies sufficient to achieve the corporate goal

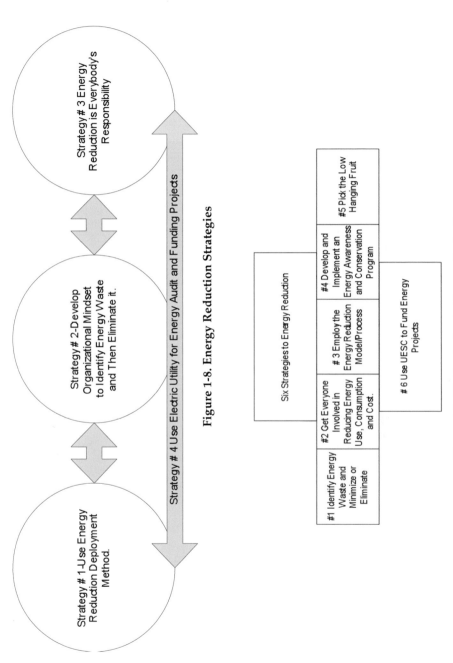

Figure 1-8. Energy Reduction Strategies

Figure 1-9. Six Strategies to Energy Reduction

and provide organizational momentum should be selected. Select the strategy or strategies that best fit your organization, then develop your plan to achieve them.

Measurement
Importance

Without measures, the energy team and management would not know how the actions taken were working or if any positive results had been obtained. The team needs measurements in every category of energy addressed to know the progress and results. Most organizations

---

1. Get all employees involved ___Yes___No
2. Identify energy waste and develop countermeasures to reduce or eliminate ___Yes___No
3. Employ the ECMS ___Yes___No
4. Follow ISO 50001 EnMS as a guide ___Yes___No
5. Develop an energy awareness training and give to all personnel ___Yes___No
6. Select the low hanging fruit and implement ___Yes___No
7. Use a UESC to fund and manage energy projects ___Yes___No
8. Perform a re-commissioning ___Yes___No
9. Perform an energy audit and develop energy plan from the results ___Yes___No
10. Implement an energy conservation program ___Yes___No
11. Implement energy centered maintenance as part of organization's PM program. ___Yes___No
12. Reduce energy consumption in our data center ___Yes___No
13. Establish a cross functional energy team ___Yes___No
14. Develop a comprehensive energy plan ___Yes___No
15. Do an energy review, develop energy performance indicators and decide a course of action ___Yes___No
16. Develop an energy champion or energy manager position and fill with the right person ___Yes___No
17. Include in the corporate strategic plan as a corporate objective to manage energy __Yes___No
18. Develop an energy awareness and conservation organizational culture. ___Yes___No
19. Other ___Yes___No

Figure 1-10. Possible Strategies

would have electricity and natural gas reduction actions, so measures for each and then their combination (rolled into one energy indicator) will be needed.

The Key Measures

The performance measures are needed to show how much electricity and natural gas were consumed each month. The units are kWh (kilowatts hours) and CF (cubic feet) or CCF (100 × CF) for electricity and natural gas respectively. These two indicators should be plotted monthly onto an Excel graph and made visible to all personnel. If desired, the units kWh and CF can both be converted into Btus (British thermal units), and combined to have a total energy indicator with Btus as the unit measure.

Figure 1-11 is a line chart and shows quickly the trend in kWh consumption at an organization. So far, FY 2014 is showing that reduction is occurring from the actions and countermeasures implemented in FY 2013 beginning in January 2013. This chart shows both progress and results for electricity reduction efforts. Sometimes, energy teams prefer column charts or bar charts to show progress. These types of charts are excellent for showing comparison between months in different years.

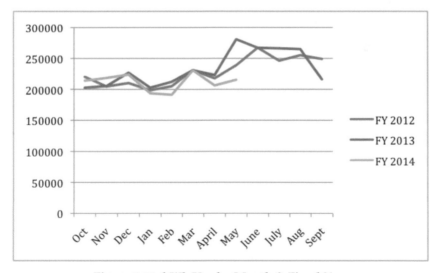

Figure 1-11. kWh Use by Month & Fiscal Year

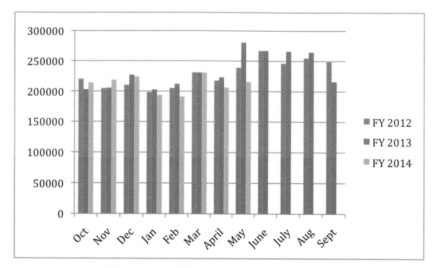

Figure 1-12. kWh Usage—Column Chart

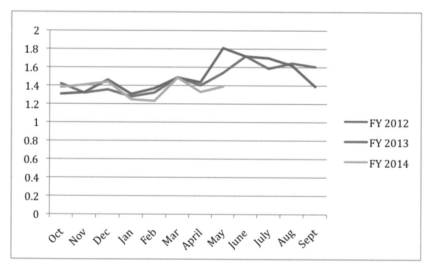

Figure 1-13. Electricity Intensity—Line Graph

To compare or benchmark with other companies or organizations, the kWh used is divided by the total square feet of the organization to get the electricity intensity. This action is called normalizing. The number of people or number of computers could be used in the denominator for normalization. However, total square feet of the facility is the traditional one used for energy—either electricity or natural gas usage.

Naturally, the electricity intensity graph shows the same trend the kWh graph did, since it is the kWh divided by 155,231 total square feet of the facility receiving and using the electricity.

In Figure 1-14, the electricity intensity is shown in a column chart. The last column is FY 2014 and shows that reductions in electricity intensity in January, February, April, and May occurred with March being about the same as the previous two years.

How did we do in reducing natural gas usage?

The line graph (Figure 1-15) shows the heavy usage in winter and almost none in summer. In January-March, the consumption was showing a reduction from last year.

The column chart (Figure 1-16) shows clearly that natural gas usage for January-March was less than the previous year. Remember, line graph for trends and column chart for comparisons.

The performance measures selected so far are:

1.   kWh usage by month.
2.   Electricity intensity kWh usage by month/total square feet.
3.   Natural gas usage by month.
4.   Natural gas usage intensity CCF/total square footage.

Now let's add two more, energy consumption by month and energy intensity by month.

To develop an energy usage indicator, one must convert both kWh and CCF into Btus. Then the Btus are added together for electricity and

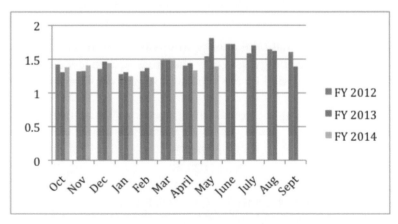

Figure 1-14. Electricity Intensity—Column Chart

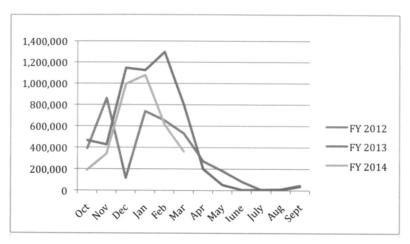

Figure 1-15. Natural Gas Usage—Line Graph

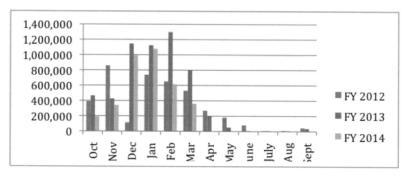

Figure 1-16. Natural Gas Usage-Column Chart

natural gas. To calculate the energy intensity for a month or year, you divide the monthly energy usage by the gross square feet and then sum of the 12 months divided by the square footage. To convert to Btus, use

1 kWh = 3412.1416 Btus   1 CF = 1020 Btus   1 CCF = 102,000 Btus
Electricity:
    FY 2013   Oct 220275   198563
    FY 2014   Oct 202838   202725

Oct 2013 = 220275 × 3412.1416 = 75,323,025 Btus
Oct 2014 = 202838 × 3412.1416 = 69,211,197 Btus

Table 1-1. kWh Usage

FY	Oct	Nov	Dec	Jan	Feb	Mar	Apr	May	June	July	Aug	Sept
2012	211613	195637	226912	190462	199575	230287	224212	244800	247838	269663	284288	223538
2013	220275	204750	210263	198563	205200	231188	217668	238950	267188	246263	255150	249075
2014	202838	205425	226913	202725								

Table 1-2. Natural Gas Usages in CCF

FY	Oct	Nov	Dec	Jan	Feb	Mar	Apr	May	June	July	Aug	Sept
2012	390,000	860,000	114,900	738,000	652,000	531,000	271,000	182,000	82,000	6,000	11,000	46,000
2013	465,000	426,000	1146000	1,124,000	1,297,000	799,000	200,000	52,000	6,000	6,000	6,000	37,000
2014	190,000	341,000	995000	1,077,000	614,000	362,000						

Natural Gas
Oct FY 2013	465,000	465,000
Oct FY 2014	190,000	190,000

Oct 2013 = 465,000 × 1020 = 47,430,000 Btus
Oct 2014 = 190000 × 1020 = 19,380,000 Btus

Next add the electricity to the natural gas, and you have one month total energy plot for Oct on the energy graph.

Oct 2013 Elect 75,323,025 + Nat. Gas 47, 430,000 = 122,753,025 energy in Btus
Oct 2014 69,211,197+ 19,380,000= 88,591,197 energy in Btus

From Oct 2013 to Oct 2014, energy reduced 122,753,025 − 88,591,197 = 34,162,828 Btus reduction.

To plot the energy graph, you will need to convert all the months into Btus and then plot onto the graph.

The kWh and CF by month indicators are converted to electricity intensity or natural gas intensity by simply dividing each month's data point by the total gross square feet. This indicator is very useful in comparing with other organizations and facilities.

Other variables are often discussed and plotted or graphed at organizations. They are shown in Figure 1-17.

Figure 1-17. KPI Characteristics

Key Performance Indicators

A performance indicator or key performance indicator (KPI) enables *performance measurement*. KPIs can evaluate the success of an organization or of a particular activity. Often success is simply the achievement of some levels of operational goal or objective. It also can be defined in terms of making progress toward strategic goals. Selecting the right KPIs depends upon a good understanding of what is important to the organization or the system or program being measured for success.

The uses of KPIs are:
- Planning—goal setting
- Trending—time series analysis
- Analysis—cross sectional analysis—several facilities comparison to prioritize
- Control—HVAC settings
- Actual versus planned efficiency and effectiveness—progress & results
- Verification—real results achieved
- Benchmarking— with other similar facilities

Why do we need to measure? Several reasons are shown below:
- A system's or program's progress or results
- Results of implementing a countermeasure
- Technical efficiency of a production process
- Energy productivity of a process
- Intensity measures to benchmark against others
- Consumption to know how we are doing

There are eight basic types of KPIs. The author added energy as a KPI. They are:
- Costs
- Consumption
- Intensity/performance
- Productivity
- Mission
- Utilization
- Activity
- Normalized
- Energy

Costs are:
- ☐ Energy costs
  - — Electricity cost
  - — Natural gas cost
  - — Fuel oil cost
  - — Steam costs
- ☐ Utility costs

Consumption KPIs are:
- ☐ kWh usage per month & by year
- ☐ CF usage per month & by year
- ☐ Pounds usage per month & by year
- ☐ Btu usage per month & by year

Intensity (sometimes called performance indicators) are:
- ☐ kWh/gross square feet
- ☐ Natural gas CF/gross square feet
- ☐ Btu/gross square feet
- ☐ Electricity costs/gross square feet
- ☐ Natural gas costs/gross square feet
- ☐ Utility costs/gross square feet

Productivity or efficiency KPIs with output/input are:
- ☐ Utility costs/student
- ☐ Energy productivity (EP) = physical output (PI)/energy input (EI) where PI is, for example, tons of steel and EI is the kWh consumption for a given time such as a year
- ☐ Energy relationship = EI/physical output which is often referred to as the technical efficiency of a particular production process
- ☐ Watts/server

Mission related KPI examples are:
- ☐ Energy costs/student
- ☐ Energy costs per credit hour earned
- ☐ MMBtu per credit hour earned

Utilization KPI examples are:
- ☐ Percent utilization of servers
- ☐ Power usage effectiveness total facility kW/data center kW

☐ Percent of our energy that is renewable

Activity KPI examples are:
☐ Percent of targeted facilities with a completed energy profile
☐ Percent of facilities with a completed walkthrough

Normalization or verification KPIs where weather and production are the two areas actually verified by comparing with the baseline are:
☐ Energy consumption (Btu) per gross square feet per degree day (DD)
☐ Energy consumption (Btu) per gross square feet per production level

Energy efficiency (EE) KPI is defined as:
☐ EE = (energy input − energy waste/energy input) × 100 where energy input is the energy consumption for the period of the measurement. Individual machines or equipment can have their EE calculated. Using the five methods recommended for identifying waste discussed later, an organization's EE can be determined.

Every selected key performance indicator should have an individual data collection plan. The data collection plan should include:

Title: _____

Frequency: _____

Type of Chart: _____

Location of Data: _____

Responsible Person: _____

Data Description
(what is included and what is not): _____

Target: _____

Summary
- Measurement is a must. Without it, the organization will not know how they are progressing and the results achieved.
- Key performance indicators must be developed for each energy source.

- Consumption by month, consumption/total gross square footage (intensity), and a summary energy indicator by month which is the sum of all consumptions by month converted to Btus is highly recommended.

- Calculating an organization's energy efficiency and developing countermeasures to reduce energy waste, thus increasing energy efficiency, is the primary thrust of energy management and reduction.

- Also include percent renewable energy as part of the total energy performance indicators. Its frequency could be quarterly, semi-annually or even yearly, depending on its degree of variability.

Teamwork and Management

The energy champion and energy team need to monitor the energy performance indicators, check the electricity and natural gas meters for accuracy and the calibration of the volt meters used to assess electricity consumption for equipment, the ECM schedule adherence, and other factors that monitoring should include. Place on a monitoring and measurement matrix, and have the energy team review periodically.

For objective and target action plans, project management should be used to ensure completion on time, within budget and with quality. All awarded projects should be controlled by project management to ensure completion on time, within budget and with quality. Verification and validation that the product and/or service contracted for was actually provided or accomplished should be a normal occurrence.

Teamwork on the energy team, walkthrough team, functional team, and any team established to assist in energy reduction is an absolute must. More on teamwork will be covered later in the book.

The Foundation
Overview

The foundation includes three areas that help ensure continuous improvement. Each of them identifies areas of opportunities. Some are deficiencies in systems implementation and others are areas where the chances of energy consumption reduction can be improved. The first, critical success factors (CSFs), are used by the energy champion and energy team to assess their progress and results and to identify areas where they can improve. After the CSFs reach a plateau, an energy reduction checklist is provided for the energy team's use. The next foundation area is self inspection or internal audit. A self-inspection checklist will be shown that can be used for the energy team conducting a self inspection or a group of other organizational personnel such as the internal audit group. Any deficiencies in what is expected to be implemented or done become areas of opportunity. The third foundation area, management review, is a proven area for generating ideas or diffusing issues leading to energy reduction opportunities. The energy champion would normally chair the management review. The energy team leader would present the inputs recommended by ISO 50001 EnMS. Others in attendance are some top management and the energy team members. An agenda with a purpose and time-limited items is followed. The outputs are normally approval of present objectives and targets and recommendation for new objectives. Self-inspection or internal audit results are reviewed as well as the energy performance indicators' (EnPIs) status. Ideas for new objectives and targets and other actions come from the review. The management review should be held at least once a year but can be more often if desired.

Critical Success Factors

The foundation first column consists of identifying and using CSFs (critical success factors) to measure progress and results through the implementation process. CSFs are those factors that must be evident or achieved to ensure project or goal achievement success. The CSFs for reducing energy are:

- Top management leadership and support
- Resources provided
- Objectives and targets established and action plans developed
- Communications
- Employee involvement
- Sufficient contributions identified to meet the corporate goal

Later, each CSF will be broken down into steps that will result as the CSF is achieved. Each step will be scored by the energy team. This will enable a total score for each month until implementation is over.

If the CSFs are achieved, then the energy goal or target will be accomplished. The CSFs can be broken down into five levels of effort, each level given a score, and then the progress of each and the overall progress and results can be measured monthly by the energy team. It can help identify where effort is needed and gauge progress as the energy team implements action plans.

I. Top Management
    1. Top management has not shown any interest or approved any energy related actions.
    2. Top management has appointed a management representative who is the energy champion.
    3. Top management has set an energy goal/objective to be achieved in three years.
    4. The energy champion has established an energy team and developed an energy policy and procedures approved by the top management.
    5. Top management participates in the management/executive reviews held annually and often gives speeches to the organization's personnel promoting energy reduction and encouraging their involvement and support.

II. Resources Provided
    1. No visible resources have been provided.
    2. Top management has agreed for personnel to participate on the energy team and the designated walkthroughs. Conference room(s) has been provided for the team meetings.
    3. Funds to develop and provide energy awareness and conservation training have been provided.
    4. Some minor projects have been funded, specifications developed, project advertised and under contact now.
    5. Both minor and major projects have contracts and are being accomplished or completed.

III. Objectives and Targets and Action Plans
    1. The cross-functional energy team has been formed.
    2. Several feasibility type objectives and targets have been established, a responsible person assigned, and action plans developed and in process of being implemented.
    3. Over five objectives and targets have been completed.
    4. Improvement projects are being scoped now and will be implemented shortly.
    5. Many improvements have been made and continue to be accomplished and maintain objectives. Targets are appearing to maintain some of the gains achieved.

IV. Communications
    1. No organizationwide communications from top management, energy champion or energy team have occurred. Some targeted communications on energy may have taken place by departments.

2. Decision by energy team to develop a communications plan has occurred and is in progress.

3. Top management has approved an energy policy, established a corporate energy reduction goal and communicated to all personnel. The communication plan has been completed and approved by the energy champion.

4. Communications on the energy plan, current energy measurements status, and progress and results achieved are being communicated at least quarterly.

5. Seldom a week passes, that some energy reduction related information is provided to all organization's personnel.

V.   Employee Involvement
1. Employees have not been asked to be involved in energy reduction effort yet.

2. Employees have received the annual energy awareness and conservation training.

3. Employees try to help and volunteer for energy related tasks when they can.

4. Employees are fully engaged in energy conservation efforts and make suggestions for further improvements or reductions when they have ideas.

5. The energy reduction program is 100% supported by the employees who contribute to its continuous improvement.

VI. Significant Contribution Has Been Identified
1. Some energy initiatives have started but it's not known if their implementation will actually reduce energy consumption and cost.

2. The energy conservation program has been developed and implemented. It has an estimated contribution of 5% energy reduction per year.

3. Facility walkthroughs have been accomplished and projects are in process of being identified.

4. The energy team is implementing ECO and developing projects with good payback (ECP).

5. Projects that once completed will enable organization to complete at least half of the targeted reduction have been funded and are under contract, the ECM has been implemented, and crafts performing the required preventative or corrective maintenance.

Figure 1-18. CSFs Measurement Instrument

The instructions and purpose shown on the CSFs progress and re-sults measuring instrument are shown below.

Instructions: Please rate each of the six CSFs by circling the 1-5 possible answers that you feel best describe the present status. The facilitator will gather your scores and total them with the other participants to obtain the total score. Each CSF has a total possible score of 1 to 5 with the 1 counting one point and the 5 counting five points. Therefore, the lowest possible score could be 6 (1 point for each CSF) and highest 30 (5 points each for six CSFs). Your score will remain confidential.

Purpose: To measure progress and results to be used to gauge implementation and identify possible future countermeasures that need to be accomplished.

When the total scores reach around 25, then the energy reduction effort should be an astounding success.

Example:
The energy team consists of six team members including the team leader. Each of the CSFs could receive a score of 1 to 5. The maximum score possible is 30 (6 members times 5 possible score for each CSF). However, each CSF total score is averaged (divide by total number of team members voting) and then all six CSFs' average scores are totaled to get the total CSF score for the month.

CSFs	May 2014 Scores	June 2014 Scores	July 2014 Scores	August 2014 Scores	Remarks
1. Top Management	2,2,1,3,2,2	2,2,2,2,2,2	3,2,3,3,2,2	3,3,3,4,3,3	
2. Resources Provided	1,1,1,1,1,1	1,2,1,2,1,1	2,1,1,1,2,1	2,2,1,2,2,2	
3. Objectives and Targets	1,1,1,1,1,1	1,1,1,1,1,1	2,2,3,2,3,2	3,3,32,3,4	
4. Communications	1,1,1,1,1,1	1,2,1,2,1,1	1,2,2,1,2,1	2,2,2,3,2,2	
5. Employee Involvement	1,1,1,1,1,1	1,1,1,1,1,1	1,1,2,1,1,1	2,2,2,2,1,2	
6. Significant Contributions	1,1,1,1,1,1	1,1,1,1,1,1	1,1,1,1,1,1	2,1,2,1,1,1	

Figure 1-19. Six Team Members' CSFs Scores May-August 2014

The average scores are shown in Figure 1-20.

The CSFs scores are plotted for each month on a line graph (Figure 1-21).

Since the program is only four months old, much progress has been made, reflected by the 13.33 total score. Much remains to be done, so it will be easy for the team to select future actions to further improve

CSFs	May 2014 Avg. Scores	June 2014 Avg. Scores	July 2014 Avg. Scores	Aug. 2014 Avg. Scores	Remarks
1. Top Management	2.0	2.0	2.5	3.17	
2. Resources Provided	1.0	1.33	1.33	1.83	
3. Objectives and Targets	1.0	1.0	2.33	3.0	
4. Communications	1.0	1.33	1.5	2.7	
5. Employee Involvement	1.0	1.0	1.17	1.83	
6. Significant Contributions	1.0	1.0	1.0	1.33	
Total	7.0	7.66	9.83	13.33	

Figure 1-20. CSFs Team Average Scores

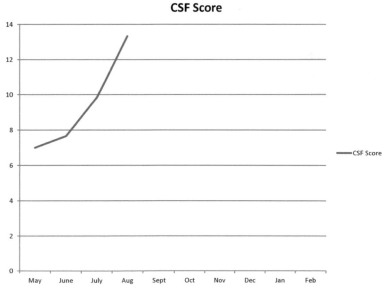

Figure 1-21. CSF Scores—Line Graph

the score. Just go into the CSF portion that is scored, and select for the near future items not done yet and target them for near term completion. Once a team gets into the 20 total score area, countermeasures to improve the score becomes more difficult. The August score shows that employee involvement and significant contributions are two areas that scored low, along with resources provided. Possible actions to improve employee involvement are to establish some additional energy functional teams or hold more management/employee brainstorming sessions. The energy team should identify more actions or projects that would contribute to meeting the organization's energy goal.

After the CSFs measurement becomes less useful when the team gets to the middle 20s in score, another measurement tool may be more helpful. It is called an energy reduction program checklist. It has 33 important items that need to be done. The energy team and energy champion should mark yes or no for each item. Each item checked yes will receive 3 points, except item number 30 receives 4 points if checked yes. Have the team and champion review this list every quarter. The items checked no, but feasible to the organization, become possible countermeasures or future plans to improve the organization's energy reduction program.

CSFs are very beneficial for about 6-7 months. Then the energy reduction checklist can be used to track progress. Have the energy team check monthly the items completed and record in the minutes. As the number increases, implementation has increased.

Figure 1-22. Energy Reduction Program Checklist

1. Has top management established a corporate energy or electricity reduction goal? ___Yes___No

2. Has top management appointed a management representative, an energy manager, or both? ___Yes___No

3. Has the management representative established a cross functional energy team? ___Yes___No

4. Has the energy team started meeting at least monthly and established their team roles and responsibilities? ___Yes___No

5. Has top management developed an energy policy and communicated it to all personnel? ___Yes___No

6. Does the energy policy include the corporate goal, management's expectations in reducing and managing energy consumption, and their

renewal energy goals? ___Yes___No

7. Have the energy champion and his appointed team completed the energy walkthroughs? ___Yes___No

8. Have the energy champion and the energy team established an energy or electricity base line? ___Yes___No

9. Have the energy performance indicators been selected and a data collection plan developed? ___Yes___No

10. Have the energy champion and energy team developed an energy plan for at least a 3-year period? ___Yes___No

11. Has the energy team established at least one objective and target and action plan and started implementation? ___Yes___No

12. Has an energy awareness and conservation plan been developed and training provided to all personnel? ___Yes___No

13. Has a communications plan been developed, approved, and followed? ___Yes___No

14. Has an emergency energy plan been developed and approved to include sufficient backup for critical operations? ___Yes___No

15. Has a procurement policy been written and included in an organizational procedure? Yes___No___

16. Has significant contribution including reducing energy waste from ECO, ECP, and ECM been accomplished? Yes___No___

17. Have the low hanging fruit actions been accomplished or at least planned? Yes___No___

18. Has an energy waste "identify and eliminate" mindset been achieved for all organizational personnel? Yes___No___

19. Were human factors and thinking used in energy planning, development, and implementation efforts? Yes___No___

20. Has ECM been implemented? Yes___No___

21. Does the organization use a computerized maintenance management system to schedule, record, and manage the ECM items completion? Yes___No___

22. Has an IT power management program been implemented at the organization for computer monitors? Yes___No___

23. Has an IT power management program at the organization been implemented for CPUs? Yes___No___

24. Has a metering plan been developed to provide essential data consumption data? Yes___No___

25. If the organization has a data center, has the energy waste been identified and plans to eliminate or minimize the energy waste been devel-

oped? Yes___No___

26. Are the key documents and records included in a centralized document control system and kept current and accessible to the organization's personnel? Yes___No___

27. Are the energy performance indicators kept current, posted or made available to all organization personnel at least quarterly? Yes___No___

28. Is the energy team using critical success factors or the energy reduction checklist to accelerate energy reduction program efforts? Yes___No___

29. Does management review the energy reduction program achievements, current objectives and targets, issues and barriers at least annually? Yes___No___

30. Is there any evidence that some processes used in reducing energy consumption and cost have been improved since the program started? Yes___No___

31. Is there any evidence of progress towards meeting the corporate goal? Yes___No___

32. Are the employees and contractors kept informed of progress and results of the energy reduction program? Yes___No___

33. Are the employees and contractors involved in energy conservation efforts? Yes___No___

A force field analysis is shown in Figure 1-23. This technique is helpful in gathering your thoughts for ideas to help sell the program to management.

Forces Pushing for	Forces pushing against
Savings/ Cost Avoidances	Additional work with no manpower increases
Top management willingness to support	Possible employee resistance to some objectives
Corporate Objective/Corporate Energy Policy	Barriers to Change and Emphasis of Status Quo
Energy Awareness and Conservation Training	People who don't know what to do to help
Cross-functional Energy team	Departments did not feel represented

Figure 1-23. A Force Field Analysis of ERD

The force field analysis shows there are strong forces pushing for the energy reduction initiative compared to the barriers or forces pushing against. This means the energy reduction initiative has an excellent chance of success.

Internal Audit and Self Inspections

Internal audits are needed to be done every 3 years to comply with ISO 50001 EnMS. If ISO 50001 EnMS is not being followed, then internal audits are not required or necessary. However, self inspections are a good driver for improvement and should be accomplished by the energy team or energy champion. The energy team should perform a self inspection once a year. Figure 1-24 is a self-inspection checklist. It is to be filled out by the energy team once a year and the team members should interview around 10 to 15 people to get their perceptions of how energy reduction and energy management are going. The self-inspection checklist can be used for internal auditing. If done, the auditors should develop an audit plan and use the checklist as the basis of the audit. In a self inspection or an internal audit, any deficiencies or non-conformances should be documented on the organization's CAR (corrective action report) and then corrected and verified that the correction was done.

The self inspection checklist enables an internal audit of actual achievements versus standards requirements in 11 major areas. They are roles and responsibilities, energy policy, energy plan, objectives and targets and their action plans, training, communications, documentation, monitoring and measurement, facility auditing and corrective actions, energy reduction deployment process and management reviews. This same checklist can be used by an organization's headquarters audit team in performing an audit of a facility or an overall organizational audit. The internal audit can be done by the headquarters internal audit department provided they develop an audit plan and assign auditors that have had some energy management training. This second-party audit needs to be accomplished every 3 years. The audit will consist of using the self-inspection checklist, interviewing personnel from top management and employees to determine their awareness, the energy champion, the energy team members, the facilities managers, and whomever else they desire to interview. They will compare what the energy reduction deployment process implementation plan said they were going to do, with what they actually did. A check on the energy performance indicators (Figure 1-24) will show what progress and results have been achieved.

Figure 1-24. Energy Audit & Self Inspection Checklist

ELEMENT 1: ROLES AND RESPONSIBILITIES

1. Have energy team member roles been identified and documented in team meeting minutes and reported to the energy champion?
   —— YES —— NO
2. Has a team leader, facilitator, and note keeper been appointed?
   —— YES —— NO
3. Do team members with identified roles understand their responsibilities, and are they being fulfilled? —— YES —— NO
4. Do the energy team members participate in meetings and attend at least 80% of all meetings? —— YES —— NO
5. Are facility staff and management aware of their responsibilities with regard to energy reduction deployment? —— YES —— NO

Comments:

_____

_____

_____

ELEMENT 2: ENERGY POLICY

1. Does the organization have an energy policy?
   —— YES —— NO
2. Has the energy policy been approved by the senior leadership?
   —— YES —— NO
3. Is the policy current and reviewed annually? (Date of last review: _____) —— YES —— NO
4. Has the policy been provided and communicated to management, employees and contractors? —— YES —— NO

Comments:

_____

_____

_____

ELEMENT 3: ENERGY PLAN

1. Does the organization have an energy plan? —— YES —— NO
2. Does it cover at least 3 years? —— YES —— NO
3. Is there a procedure in place to ensure the plan is periodically updated and when a major revision is in order? —— YES —— NO
4. Are the energy policy and the corporate goal included in the energy plan?
   —— YES —— NO

5.   Are the intended objectives, actions, and projects of the energy goal clear?
                    —— YES —— NO

Comments:

_____

_____

_____

ELEMENT 4: OBJECTIVES & TARGETS AND
ENERGY ACTION PLANS (EAP)

1.   Has the energy team worked on or initiated at least one objective and target this year?   —— YES   —— NO

2.   Have objectives and targets and action plans been documented?
                —— YES   —— NO

3.   Do objectives and targets address the respective facility energy efficiencies or deficiencies?   —— YES   —— NO

4.   Has a documented action plan designating responsibility and detailing steps to completion been put in place for each objective and target, and written to describe how the team will successfully meet the target?
                —— YES   —— NO

5.   Are objectives being reviewed for progress at the team meetings?
                —— YES   —— NO

6.   Is the status of each action plan reflected using a stop light symbol at each meeting? (Date of last review: _____)
                —— YES   —— NO

Comments:

_____

_____

_____

ELEMENT 5: TRAINING

1.   Has energy awareness training been developed and presented to middle management, employees and contractors?   —— YES —— NO

2.   Have the *Training Needs* been reviewed within the past year?
(Date of last review:_____)   —— YES   —— NO

3.   Have employees received energy conservation training?
                —— YES   —— NO

4.   Have management, energy champion, and energy team received energy reduction deployment training?   —— YES   —— NO

Comments:

_____

_____

_____

ELEMENT 6: COMMUNICATION

1. Does the energy champion and energy team communicate relevant information to the facility middle management, employees and contractors?
                                                            —— YES —— NO
2. Do employees know to whom to communicate energy concerns?
                                                            —— YES —— NO
3. Is there a communications plan?   —— YES —— NO
4. Are external communications adequately addressed?   —— YES —— NO

Comments:

_____

_____

_____

ELEMENT 7: MONITORING AND MEASUREMENT

1. Have significant electricity users that require monitoring or measurement been identified in the monitor and measure plan?   —— YES —— NO
2. Has the monitor or measure plan been reviewed within the past year? (Date of last review:_____)   —— YES —— NO
3. Is the required monitoring and measuring being conducted for the energy performance indicators?   —— YES —— NO
4. Is the monitoring/measurement relevant to the projects under construction being accomplished?   —— YES —— NO

Comments:

_____

_____

_____

ELEMENT 8: DOCUMENTATION AND CONTROL OF DOCUMENTS

1. Is all required documentation filed on the facility's energy document control site?   ___ YES ___ NO
2. Is all documentation filed in the correct folders according to the energy documentation guide procedure?   ___ YES ___ NO

3.   Is all documentation on the document control site current?

                              ___ YES ___ NO

4.   Are the documents used by the energy team the current energy program documents?   ___ YES ___ NO

5.   Does the energy documentation provide direction and instructions to other related documents, records, reports, schedules and registers?

                              ___ YES ___ NO

Comments:

_____

_____

_____

ELEMENT 9: FACILITY AUDITING AND CORRECTIVE ACTION

1.   Have self-inspections been completed annually?   — YES   — NO

2.   Were all non-conformances identified and corrective action taken to address the non-conformance?   ___ YES   ___ NO ___ N/A

3.   Were all CARs from the previous second-party audits resolved appropriately?   ___ YES   ___ NO ___ N/A

Comments:

_____

_____

_____

ELEMENT 10: MANAGEMENT REVIEW

1.   Has the management review been conducted at least annually? (Give the date(s) of last review: _____)

                              — YES   — NO

2.   Is the head of the facility, or their designee, aware of the status of the energy reduction program?   — YES   — NO

3.   Does the head of the facility, or their designee, provide guidance and direction for the energy reduction deployment?   — YES   — NO

4.   Does the management review cover the appropriate elements including both inputs and outputs?   — YES   — NO

Comments:

_____

_____

_____

ELEMENT 11: ENERGY REDUCTION DEPLOYMENT

1.   Were the walkthroughs successful in identifying the energy
      information necessary for ECP and ECM?   —— YES   —— NO
2.   Has the ECM been placed into a CMMS?   —— YES   —— NO
3.   Has CMMS scheduled and accomplished the new preventative
      maintenance items?   —— YES   —— NO
4.   Have good payback projects been identified to reduce energy use?
                                                          —— YES   —— NO
5.   Has the energy team identified significant contribution to achieve the
      corporate objective?   —— YES   —— NO

Comments:

_____

_____

_____

Management Reviews

A management review should be held at least annually and chaired by the energy champion and attended by other executives, energy team, and guests.

Inputs should include an agenda using PAL, a PowerPoint presentation, and a sign-in sheet to document attendees. Any self inspections, internal audits, CARs, performance indicators status, issues, complaints, and barriers should be discussed.

Outputs should be recommendations for improvements, new objectives and targets, approval of plans such as an emergency plan, communications plan, and an energy procurement plan, training plan and action items to be completed.

The Components

Introduction

ISO 50001 Energy Management System (EnMS) is an excellent system for managing energy for any organization. However, it does not provide a clear deployment system like policy deployment or strategic planning does. If an organization decided to reduce their energy by 10% or 20% in two years, the deployment actions to do so are not spelled out clearly in the standard. The primary purpose of this book is to provide a clear deployment process that will enable an organization to reduce energy using the appropriate components and activities. Figure 1-25 shows the major components of the process.

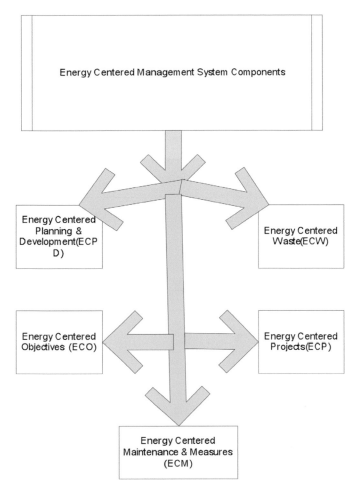

Figure 1-25. Major Components of Energy Reduction Deployment

The components start with planning and developing the deployment (ECP&D), the identification of energy waste (ECW), the energy team establishing objectives and targets to reduce energy (ECO), the identification of energy reduction projects, their funding, and project management (ECP). And the implementation of preventative and corrective action items that reduce energy use (ECM). Due to the amount of material that may be required to adequately describe each of these, it is best to cover each one in a separate chapter. In expanded form, the energy centered management system is diagrammed in Figure 1-26, and will be explained in the next chapters.

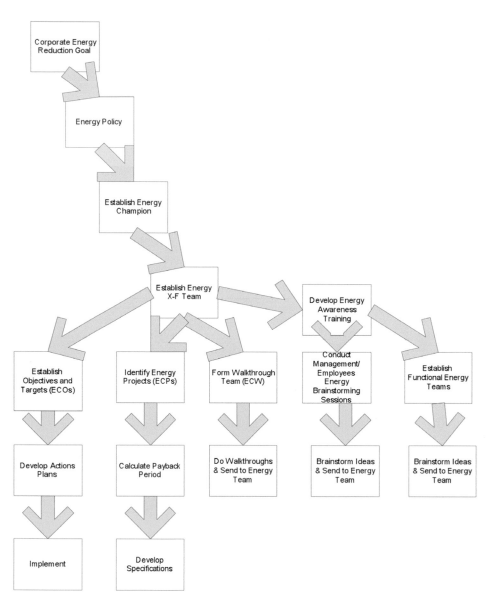

Figure 1-26. Energy Centered Management System

Chapter 2

Energy Centered Planning and Development (ECP&D)

PURPOSE

The primary purpose of ECP&D is to get the deployment started through engaging top management, and gaining their commitment and support. Top management commitment is shown by establishing an organizational goal to reduce energy use and consumption. The goal should be made into a corporate objective by stating how much reduction is to be accomplished and by what date. Next, the management should appoint one of its own, a management representative. He or she will function as the energy champion. If the organization already has an energy manager, he or she may be the energy champion representing the organization. The energy champion's first duty is to develop an energy policy and get management's approval. Next, he or she should establish a cross-functional energy team with at least one member from facilities or engineering. Together the energy champion and energy team need to develop an energy plan that outlines what needs to be done, by whom, and when, to meet the corporate objective and goal. The total elements in ECP&D are shown in Figure 2-1.

An Environmental Scan—A Beneficial Pre-deployment Technique
The purpose of the environmental scan is to simply gather the information from both internal and external sources. It is not to make any decisions at this point as to what to do with the data. That step comes next, in the SWOT analysis. Environmental scanning is a process that systematically surveys and interprets relevant data to identify external opportunities and threats that may affect or impact an organization. An organization gathers information about the external world, its competitors and itself (internal scanning). The company can use the information gathered by developing or revising its vision and mission, and by changing its strategies when management deems it should.

Figure 2-1. Energy Centered Planning and Development (ECP&D)

The environmental scanning process consists of performing several steps. The first step is for an organization to gather information about the world in which it operates. That includes information about the economy, government, laws and demographic factors such as population size and distribution.

Now, the organization should focus on its competitors. The organization should examine the research for trends, opportunities and threats that might impact its business.

The next step is to conduct an internal scan of the organization. Examine the organization's strengths and weaknesses. Consider where the organization is now and where it plans to be in 5 or 10 years. Interview or survey leaders of the company or organization.

When conducting an environmental scan, a variety of methods are available to collect data. They are reviewing publications, conducting focus groups, interviewing top management inside and leaders outside the organization, listening to media, attending civic associations and research at the library. After the data are collected, the final step is to analyze the data, turn them into information, then identify changes that should be made.

A team approach works best in most organizations. First, having team members from various parts of the organization ensures that functional areas, services, or issues are not overlooked and are included. Next, having a team helps create buy-in to the whole organization and allows for transparency of the process. Transparency helps build trust and buy-in in the energy reduction deployment plan, which is an enabler for its success.

The environmental scan team members will be unique to the organization, but typically the team should consist of frontline supervisors, representative employees across the organization, and mid-level managers. The team approach includes taking these steps:

- Each team member researches one or more external or internal factors.

- Each team member prepares a list of relevant factual data that supports the factors that they have researched.

- Members share their completed lists with the entire team so that each member has a complete list of all factors.

External Area	Type of Information	Source
Demographic Information	Qualified energy auditors and energy consultants services available for a cost	Internet and Telephone Books
Government Influence	Encourages renewal energy use. Provides rebates on renewal energy applications.	Federal Regulations, Executive Orders, State Statutes
Economic Conditions	Economy growing slowly and energy cost is around 30% of company's cost.	Internet
Geographical Information	Reliability of energy delivery systems vary from region to region.	Internet, Utility Publications
Technology Information	More energy friendly equipment available.	Trade Publications & Internet

Figure 2-2a. Energy Environmental Scan—External Area

Internal Area	Type of Information	Source
Unions Information	Trade Unions. No major issues at this time.	Union Steward, Human Resources
Employees Information	Turnover rate is very low. Employee satisfaction is high.	Human resources and survey sent to all employees.
Budget Information	Energy cost is 30% of budget and increasing 3 to 5 % a year.	Checked budget data and reports. Interviewed Comptroller and accounting supervisor.
Political Information	Government climate is positive for reducing energy use and consumption. Top management is in favor of doing this.	Interviewed a cross section of top management.

Figure 2-2b. Energy Environmental Scan—Internal Area

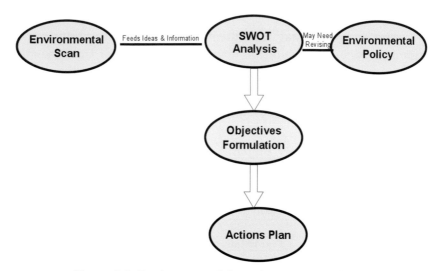

Figure 2-3. Environmental Scan & SWOT Analysis

A SWOT Analysis

A SWOT analysis is a structured planning method that enables evaluation of the strengths, weaknesses, opportunities, and threats involved in a project or in a business initiative such as reducing energy consumption. A SWOT analysis can be carried out for a product, place, industry, an initiative or even a person. It involves defining the objective of the business venture or project and then identifying the internal and external factors that are favorable. Establishing an objective should be done after the SWOT analysis has been performed. However, the goal of reducing energy consumption is a natural, and the SWOT will help identify specific sub-objectives that could prove beneficial. This technique facilitates or enables achievable goals or objectives for the organization.

• **Strengths**: characteristics of the business initiative that give it an advantage over others.

• **Weaknesses**: characteristics that place the business initiative at a disadvantage relative to others.

• **Opportunities**: elements that the initiative could exploit to its advantage.

• **Threats**: elements in the environment that could cause trouble for the business initiative.

Identification of SWOTs is important because they can identify later steps in planning to achieve the objective.

Strengths

1. Energy reduction saves money.

2. Organization has a lot of past electricity and natural gas consumption data for analysis.

3. Facilities personnel have some energy management training and experience.

4. Some electric utilities do energy audits and assist in funding projects with excellent payback.

5. Facility has an electric meter and a natural gas meter presently.

6. Reducing energy consumption and cost lends itself well to be a corporate objective that deployment can involve everyone in the organization.

7. There is a lot of information on the internet on reducing energy use and consumption.

8. ISO has completed a standard called ISO 50001 Energy Management System (EnMS) that can serve as a guide or actually be implemented if management desires.

Weaknesses

1. There are no full time energy management personnel assigned to the organization.

2. There has been no past improvement effort of this magnitude attempted by the organization before.

3. Only part time personnel are available for the energy teams that will be needed.

4. There is no training available on energy management presently.

5. Strategic planning has not been used by the organization on a regular basis.

Opportunities

1. A chance to save money for other important uses.

2.   Organization can get favorable publicity for its efforts by the stockholders.

3.   Gives organization a chance to contribute to the environment by saving energy.

4.   Can get assistance form electric utility and upgrade some of the old air conditioning and heating equipment.

5.   Can get all the organization's personnel working and contributing together for a worthwhile effort.

Threats

1.   Failure due to organization's limited experience in such a large initiative.

2.   Insufficient supply of electricity available during high energy use days.

3.   Electricity and/or natural gas cost could significantly increase due to unplanned situations such as military conflict, storms, etc.

Objectives and targets usually come from the SWOT analysis. An example would be "Develop an energy emergency plan by Dec. 31, 2014."

Strengths	Opportunities	Helps?	Other
1. Energy Reduction saves money.	A chance to save money that can be used for other mission requirements.	Yes. Can use this to help sell top management on reducing energy.	Includes a Force Field Analysis, derived from results of SWOT, which can help sell management also.
2. Facilities personnel have some energy training and experience.	Gives organization a chance to contribute to the environment by saving energy.	Yes. Includes facilities personnel on energy team and walk through team.	Energy Team will need other cross-functional members.
3. Electric Utility does energy audits and assists in funding energy projects with excellent pay back.	Can get assistance from electric utility and upgrade some of the old air conditioning and heating equipment.	Yes. Later in energy reduction deployment, ECP projects can be identified scoped, and payback period calculated.	After the low hanging fruit are picked and other plans developed, ECP will be addressed.

Figure 2-4a. SWOT Analysis—Strengths—Opportunities

Strengths	Weaknesses	Helps?	Other
Reducing energy consumption and cost lends itself well to be a corporate objective that deployment can involve everyone in the organization. 2. ISO has completed a standard called ISO 50001 Energy Management System (EnMS) that can serve as a guide or actually be implemented if management desires.	There has been no past improvement effort of this magnitude attempted by the organization before. 2. Strategic planning has not been used by the organization on a regular basis.	Yes. Even though the organization is inexperienced in this area, this is an excellent initiative to gain the experience and be successful in saving money.	Possibly an external energy facilitator would be helpful and worthwhile.

Figure 2-4b. SWOT Analysis—Opportunities—Threats

Strengths	Threats	Helps?	Other
1. Facilities personnel have some energy management training and experience.	1. Insufficient supply of electricity available during high energy use days.	Yes. Facilities personnel can help plan the backup supply needed for critical activities.	Energy Team with facilities and engineering help should develop an emergency plan that will solve this problem.
2. Energy Reduction saves money.	2. Electricity and/or natural gas cost could significantly increase due to unplanned situations such as military conflict, storms, etc.	Yes. Means more savings provided the energy supply is available for the organization's demand.	Not very probable to happen, but has in the past.

Figure 2-4c. SWOT Analysis—Strengths—Threats

ENERGY POLICY

The energy champion with the help of the energy team develops a draft energy policy and present to top management. Top management approves it if it is compelling, states what they desire to happen and their energy reduction goal and renewable energy goal. The two can be interrelated, for example, if you change the hot water heaters from electricity or natural gas to solar, it will reduce the energy consumption and increase renewable energy use. Once the energy policy is developed, it is communicated to everyone in the organization and posted on bulletin boards throughout the organization. The energy policy is similar to the vision in strategic planning. ECP&D is the first stage of any energy use and consumption reduction initiative.

An energy policy serves as a formal declaration that your organization plans to pursue energy management and reduction as a corporate priority. It establishes energy management and reduction as a core value and a means to achieving significant cost savings. It signals to employees, investors, shareholders, owners, tenants, and the public that your organization is focusing on energy management and reduction, and it describes how committed you are in achieving the energy policy.

Why is an energy policy needed? It is similar to a vision for other strategic improvements.

1. Any organization is likely to be more successful in reducing if there is a clear statement of the goal or corporate objective.

2. The organization will understand and appreciate the value of energy management more if it is able to measure performance against an established goal and/or target.

3. Energy management and reduction activities will be more effective if an energy champion and energy team are established and the program is adequately funded.

4. Energy management is more likely to be accepted and supported throughout the organization if it has formal support and blessing from top management.

An energy policy should include:
•   Any commitments your organization will make in managing energy. For example:

"We are committed to purchasing and consuming energy in the most efficient and environmentally responsible way possible and at least cost. To achieve this we will:

•   Improve energy efficiency continuously by implementing effective energy management objectives and targets and energy projects that support all operations and stakeholders' satisfaction while providing a safe, comfortable, and inspiring work environment.

- Become a best-in-class energy-efficient organization in our industry on an electricity intensity (kilowatt-hour per square feet) basis."

- The energy policy should be clear, concise, and written at a high level. It should include broad energy objectives and targets that apply to your organization, but not include the detailed actions and activities, such as your energy implementation plan. It should give a time frame for accomplishment."

An Example follows:

ABC Company's corporate mission is to provide:
- The best possible facilities and an excellent work environment.
- Maximum value to our shareholders

In pursuit of this, ABC Company will strive to achieve a best-in-class reputation for energy management. We are committed to becoming a high performance company that uses energy in the most efficient, cost-effective, and environmentally responsible manner possible.

Energy management will play a key role in our business. It will support our plan to maximize profitability, strengthen our competitive position, and provide our customers with excellent customer service. Our program to reduce energy consumption and prevent pollution will also support our commitment to our employees, the environment, and our community.

Toward this end, ABC Company shall work towards continuously improving energy performance. We will develop an energy plan that will ensure energy reduction of 20% by year end 2016.

Signed: CEO or Organization Head   _____ DATE:

What else may be included if desired by management? They are:
- Organization's strategies.

- Energy champion's name.

- Top management expectations of the support it desires from supervisors, employees, and contractors.

- Multi-years goals or targets.

- Objectives of the energy reduction deployment effort.

- Mission statement.

- Green house gas reduction & renewal energy goals.

- Review procedures.

- Description of organization to manage energy.

- Assigned responsibilities of teams, committees or arrangements.

- Show management's commitment to providing resourcess to energy reduction, and to contnuous improvements.

ENERGY PLAN

An energy plan needs to be developed to:
1. Guide the organization's energy reduction efforts.
2. To communicate the energy program to all organization's personnel and encourage their support.
3. To outline the actions and activities necessary to achieve the organization's energy objective and goal.

The energy team needs to develop a draft energy plan and brief the energy champion. Once the energy champion approves the energy plan, then it is briefed to top management to get their approval and support.

A possible table of contents would include

I. Energy plan purpose and Energy Centered Management System (ECMS) explanation

II. Energy policy

III. Corporate energy goal(s) and objective(s)

IV. Energy champion roles and responsibilities

V. Energy team charter & roles and responsibilities

VI. Objectives and targets

VII. Action plans by next five years include energy efficiencies and renewable energy projects

The action plan for the first year will include the energy walkthrough identifying waste and proposed preventative maintenance, projects, and metering. Also, development of an energy conservation plan and its communication to all personnel should be a first year activity. This can be accomplished by an objective and target (O&T) under the ECO and then transferred to ECM for implementation. If an energy audit is recommended or re-commissioning than this should be placed in the appropriate year.

APPOINTING A MANAGEMENT REPRESENTATIVE

Top management should appoint a management representative to handle the day to day energy matters. He or she will be called the energy champion. The duties will be:

☐ Develop energy team charter

☐ Establish an energy cross functional team

☐ Keep top management informed and provide resources

☐ Assist energy team in overcoming barriers

☐ Lead development and approval of a corporate energy policy and communicate it to all

☐ Encourage everyone's support & involvement

☐ Conduct management reviews

A team charter paves the way for success by providing clarity of what the purpose is to having the team. The team type such as functional, cross functional, task team, strategic team or a combination such as an energy cross functional team may be mentioned. The corporate goal or goals (energy reduction, increase renewable energy percentage, may be included plus any special requirements such as time period the purpose should be completed. Also, communications and/or report requirements may be outlined.

ESTABLISHING AN ENERGY TEAM

The energy champion should establish a cross functional energy team. The team should consist of five to ten people, the size being determined by the need to have all major departments represented on the team. At least one or two members should be from the facilities and/or engineering departments.

The team roles and responsibilities are:

- Team leader-appointed by energy champion or elected by the team

- Facilitator—outside or inside facilitator

- Document control manager—a volunteer

- Note taker—a volunteer

The following factors must be considered in establishing any team.

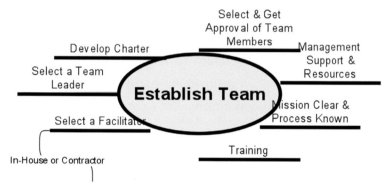

Figure 2-5. Establish Team

Approval from the department heads for selected team members to be on the energy team should be done by the energy champion. If a facilitator is experienced, he or she could give training to the team on the process that will be using and some of the tools and techniques that will be employed. The discussion of the charter by the energy champion should be part of the training or be covered at the first official team meeting.

The characteristics of a high performance team are shown in Figure 2-6.

Trust and Support—Team members have mutual trust in each other and in the team's assignment or purpose—each member feels free to express their feelings, suggestions, and ideas.

Communications—The team engages in open discussions, and

Figure 2-6. Characteristics of a High Performance Team

everyone gets a turn to participate and contribute—from the ones who talk a lot to the silent ones. Conflict is managed and actually contributes to clarifying issues and opinions. The energy champion or top management are kept informed of team actions, barriers, issues, progress, and results.

FOCUS—Ground rules are established for team meetings.

A process is selected that when followed will lead to achieving the team purpose.

An agenda is developed for each meeting to include P—Purpose, A—Agenda, and L—Limited. The agenda is sent to team members prior to meeting.

Everyone understands both team and individual performance goals and knows what is expected.

Commitment and Support—All team members are committed to the team goals and to accomplishing the team purpose.

Once a decision is made, each team member supports the decision even if he or she did not vote for it.

Every team member is working toward the same goal(s).

Accountability—Each team member carries his or her own weight and respects the team processes and other members.

All team members accomplish their assignments, including homework outside of the meeting.

Everyone is responsible and accountable for their actions.

Solutions and Results—The team meets their purpose in a very short time with quality solutions and results.

The team is recognized as a high performance team and wants to stay together as a team after their success(es).

ESTABLISHING ENERGY PERFORMANCE
INDICATORS AND A BASELINE

The energy team analyzes the energy bills from the past to determine trends and establish a baseline consisting of one representative year of each energy type by month that is typical of the organization's consumption. Most of the time, it is the last year's consumption by month. It does not have to be, if something unusual happened that made the consumption not normal. Any year in the last 3 years can be the baseline or the monthly average for the 3 years. The baseline is very useful in that future consumption can be compared to each month to verify that improvement (reduction) has occurred or been achieved. The energy champion and energy team will develop energy performance indicators (EnPIs) for each energy type used. Normally, electricity and natural gas are the primary energy types used. The consumption by month is an important key indicator. It can be trended using a line graph versus the baseline year to show reduction. The units are kWh for electricity and CF (cubic feet) for natural gas. To benchmark the organization with others, an intensity indicator is developed by month. It is simply the consumption by month divided by the total gross square feet of the facility. The denominator could be computers or people, but for the energy industry, it is total gross square feet. Once a quarter, all the energy indicators are converted to make one energy indicator showing total Btus consumed. For renewal energy the most useful indicator is the percent of total energy that is renewable. Since this indicator does not change frequently, it is normally calculated only once a year.

To fully understand the above energy performance indicators and others, it is important to know where indicators come from or originate. Measures or indicators come from either objectives or processes. The electricity cycle is shown in Figure 2-7. It does not include, but could, electricity generation and transmission.

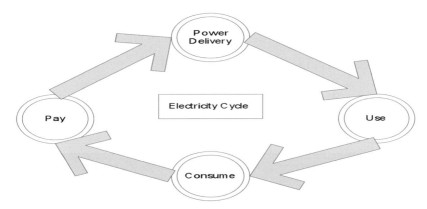

Figure 2-7. Electricity Cycle

Power Delivery—Our objectives are:
1. At a low price
2. Excellent reliability
3. Excellent quality
4. No power factor—reactive power
5. Includes some renewable energy
6. Demand response available

Use—Our objectives are:
1. Use where cost-effective
2. Use alternative where not cost effective (natural gas for heaters, solar for hot water)
3. Know the significant energy users (SEUS) and put in place operational controls and sub-meters.

Consumption—Our objectives are:
1. Consume as little as possible.
    — Practice conservation
    — Have an energy awareness work culture
    — Organization and deployment reduction system being practiced to reduce energy consumption
2. Measure consumption
3. Track, monitor, and evaluate energy performance indicators (EnPIs)

Pay—Our objectives are:
1. Understand electric bill
2. Ensure no overcharge
3. Use electric bill to provide data for EnPIs

What are the indicators that measure the objectives? They are shown in Figure 2-8.

Electricity Cycle Phase	Metrics/Measures
Power Delivery	✓ Cost (Cents per KWH), Service Unavailability, # Of Service Interruptions, Average Duration Time, # Momentaries, Power Factor, Per Cent Renewable Energy
Use	Per Cent SEUs that have Operational Controls and Per Cent SEUs that have own meter or sub-meter.
Consumption	✓ KWH Consumption, Energy Intensity, ELF
Pay	✓ Cost, Demand Response Dollars, Power Factor Adjustment

Figure 2-8. Electricity Cycle Measures

The indicators with a check mark should be:

*   Developed and follow the data collection plans for each major energy performance indicator (EnPI).

*   Kept EnPIs current, visible to all, and easy to read.

*   The energy champion and energy team monitored and initiated corrective actions when needed.

*   The indicators enabled the "check" phase of the plan-do-check-act cycle.

Energy Performance Measures

- Will vary a little by organizations depending on their energy uses.

- All should have electricity kWh consumption by month, CF or CCF natural gas consumption by month, changed to Btus, and both converted to intensity by dividing by the total gross square footage.

- Both electricity and natural gas consumption should be converted to Btus and then summed monthly to develop a total energy consumption performance indicator. Monitoring electricity and other energy costs is useful.

- Calculate at least quarterly the electricity load factor (ELF).

- A percent renewal energy indicator is helpful if it changes some in a year.

The electricity and natural gas intensity indicators should be converted to Btus and combined prior to benchmarking with other organizations on their energy intensity. Figure 2-9 gives the conversion to Btus factors.

Fuel BTUs	Other Units
Electricity 3,412 BTUs=	1 KWH
Natural Gas 1,027 BTUs=	1 Cubic Feet (CF)
Fuel Oil 138,690 BTUs=	1 Gallon
Hot Water 1,000 BTUs=	1 Pound of Steam

Figure 2-9. Btus Conversions

The demand period is the time interval during which the flow of electricity is measured (15-, 30-, or 60-minute increments, depending on the rate structure and type of meter). Usually peak demand is the highest demand over the demand period in the billing period (normally a calendar month). In other words, it is the highest amount of power your facility requires at a given time. Often this leads to demand charges. You can think of demand charges as overhead expenses that your utility incurs to provide the electricity infrastructure that is capable of meeting your largest electric load.

- Electric load factor (ELF) is an indicator that shows if peak demand is high for your facility. It is an indicator of how steady an electrical load is over time. The optimum load factor is 1 or 100%. The closer to zero, the more you are paying for electricity.

- ELF (%) = Total kWhs/#days in electricity bill cycle × 24 hours/ day/peak kW demand.

In other words, it is the average demand/peak demand for a given period of time. From your electricity bill, get the kWhs used and the peak kW. Next look for days included in the bill. Multiply these days by 24 (hours per day). Divide this number into the kWhs. Then divide what you get by the peak kW. Multiply this number by 100 to get the percentage.

- For example: If a facility used 125,000 kWh in July where the billing period covered 30 days, the peak kW demand was 218, and the ELF is calculated by (125,000/30 × 24/218) × 100 = 79.64%.

- Action should be initiated to increase load factor when you are 60% or lower. If low, shift electricity intensity processes to other times. By increasing load factor, you will reduce the impact of monthly demand (kW) charged on your electric bill.

- This calculation only needs to be done quarterly to ensure ELF has not significantly changed.

- Goal—ELF 60% or higher

How can we make that happen through favorable actions?
1. Increase kWh consumption—Would increase ELF, but not cost effective.

2. Decrease days in bill period—Would increase ELF but under control of utility only.

3. Decrease demand kW—Yes, this is what is needed.

How can the peak load be decreased? 1. Reduce load (shed load) or 2. Increase capacity through on-site generation.

- Sheddable loads are those that can be easily turned off and restarted, without serious impact on processes or staff.

- Typical sheddable loads include HVAC systems, non-sensitive industrial machines, some lighting, appliances, and some computers.

Loads can be prioritized by the amount of time they can be turned off before they impact productivity, safety, or profitability.

For example, a refrigeration unit may be able to be shut down for up to 20 minutes, but must receive electricity within that timeframe, while lighting in a tool or supply storage area may be sheddable for as much as 60-90 minutes.

Loads can be categorized for automated shedding.

• Critical—Important processes related to safety.

• Essential—Any shedding will reduce productivity, production and/or profit.

• Non-essential—These are the loads to shed.

Use power generators—Purchased to meet critical and all or some of the essential load.

— Will need environmental permits if underground, and environmental conditions if above ground.

— Emission standards must be complied with for the year.

Renewable energy consisting of **solar**, **wind**, **hydroelectric**, **geothermal**, and **biomass** should be graphed and communicated at least yearly.

A baseline should be established for each energy performance indicator. Each major energy performance indicator should have a

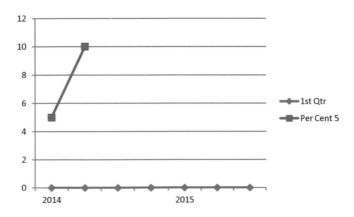

Figure 2-10. Renewable Energy Percent.

baseline so progress and results can be compared in the future to verify that improvement has been achieved. Normal procedure is to take 2 or 3 years of data and through analysis determine a year most representative of energy use, and that will become the baseline year. It does not have to be last year, but should be within a 3-year timeline or sooner, if feasible.

CORPORATE PROCEDURES

If the energy champion sees the deployment lasting several years, or if it becomes a continuous journey as required by ISO 50001 EnMS, may task the energy team to develop corporate procedures to guide the deployment, and if there is more than one facility involved to ensure consistency in the implementation.

In this phase, the energy team starts the energy plan showing the organization's plan to reduce energy at least 3 years forward. The phase or component, energy centered waste, will need to feed the energy team their findings for the plan to be finalized.

ENERGY TRAINING

Mandatory Training

Top management is recommended to receive *Energy Reduction Deployment Model & Process*.

All personnel in the organization should receive two mandatory training courses: *Energy Awareness* and *Energy Conservation*.

Competence Training

Recommended to the people involved in these processes:
— Reading utility bills
— Reading meters
— Advanced metering
— Power factor
— Using BAS and
— Walkthrough process including using electricity measuring devices.

When personnel in the energy reduction process are unfamiliar with certain important areas, training should be provided.

Chapter 3

Energy Centered Waste (ECW)

IDENTIFYING ENERGY WASTE

First, energy waste needs to be identified. Only then can actions be taken to achieve the energy consumption reduction goal. The entire organization should develop the mindset to identify energy waste. Until it is identified, no fix can be applied. Developing a walkthrough team, having a kickoff meeting, performing the walkthrough(s), documenting and evaluating the findings, recommending to the energy team which component (ECO, ECP, ECM) they should be assigned and what should be included in the energy plan are the walkthrough team's mission. Energy is wasted in a lot of different ways. The present energy waste modes or types identified are:

- Running when it should be turned off (inoperable controls or no control)
- Having to work harder
- Worn
- Winding failure
- Insulation failure
- Breaking, bending, jamming, and bond failure
- Leaks
- Stuck in open or closed position
- Under performing
- Over performing
- Overheats
- Waste build-up
- Incorrect settings

Figure 3-1. Energy Waste Types

An example of each energy waste type is shown in Figure 3-2.

---

1. Running when it should be turned off (inoperable controls or no control)— motors, computers, monitors
2. Having to work harder—HVAC with dirty filters
3- Worn—fans & bearings
4. Leaks—windows, doors, air compressors
5. Winding failure—motors, pumps, and fans
6. Insulation failure—motors, pumps, and fans
7. Breaking, bending, jamming, and bond failure—motors, pumps, and fans
8. Stuck in open or closed position—economizers
9. Under-performing—dirty or grease on lights
10. Over performing (copiers, printers)
11. Overheats—condensers
12. Waste build-up—in boilers
13. Incorrect settings—exhaust fans

Figure 3-2. Energy Waste Types with Examples

A question should be asked during the walkthrough if any item or equipment is likely to waste energy like any of the energy waste types. Energy waste is when equipment, building infrastructure or envelope, or electronics uses excess energy due to some problem being experienced, but it is still able to perform its function.

To the energy waste diagram originally designed by the author, three additional ways to identify the waste have been added. The first

Figure 3-3. Energy Centered Waste (ECW)—Identify Waste

is for management to hold employee brainstorming sessions. At these sessions, consisting of several small functional areas or a large functional area, energy awareness training is first given and then employees brainstorm ways that energy can be reduced in their work areas. These ideas are given to the energy team that includes those reponsible for items on the energy waste list and for the energy conservation program and training.

Second is to form functional teams or natural working groups to address energy waste identification in their work area. Their ideas are given to the energy team to help build the energy conservation program, energy plan, and training. In the energy awareness training, let everyone know who the energy champion and energy team members are, and encourage all personnel in the organization to contact one of them or put into the organization's suggestion box or program, any energy related ideas, concepts, issues or problems.

The third area is to encourage all personnel to use the organization's or company's suggestion program and submit suggestions about energy waste, possible objectives and projects.

ENERGY WALKTHROUGH(S)

A qualified energy walkthrough team is established to identify energy waste and make recommendations as to the fix. The team should include the energy manager, facilities manager, facility engineer, one member of each craft, plus any other person the energy champion thinks can contribute. Normally instead of just one walkthrough, there are several focused ones done. Typically, they are as seen in Table 3-1.

Whether it is electricity, natural gas, steam, wind, solar or whatever, the energy waste will be identified. The leaks around doors, overbuilding of equipment such as too large of cooling towers, inoperable controls and items wasting energy due to poor or inadequate maintenance. The team needs to find the significant energy users and locate where energy waste is occurring or where there is a high probability of its occurring. They should identify where improved maintenance will help or if projects such as replacing air conditioners or boilers or lights would be beneficial and cost effective. The team members are often called energy use busters, energy waste eliminators, or other appropriate names. The success of the energy reduction deployment will

Table 3-1. Focused Walkthroughs

Walkthrough	Observations
1. Occupancy Sensors	Observe infrequent visited areas and determine whether an Occupancy Sensor will save energy. Look at rest rooms, break rooms, copying or printing areas, mechanical areas, hallways and other areas.
2. Lights in administrative areas.	Note type such as T-12s, T-8s, & T-5s. Look for areas day lighting can be used and skylights would help. Look at light bulbs and see if they are dirty with film covering them.
3. Building Envelope	Look for leaks in doors and windows. Determine if windows should be glazed, caulked or replaced. Weather strip the doors where needed or replace them.
4. Walls and Roof insulation	Check the insulation level and determine if more would help.
5. Motors and other equipment except HVAC	Note each one and check the switches and sensors associated with each.
6. Data Centers	Look for hot and cold aisles and whether hot air is kept from commingling with the cold air on its return to the computer (CRAC).
7. Security Lights	Check to see if they are adequate and energy friendly.
8. HVAC	Note brand, capacity, date installed, the motors and switches associated with the system, check roof vents, and other parts for adequacy and maintenance.
9. Building Automation System(BAS) and metering	See if BAS is outdated. Note where additional metering can help identify potential problem areas.
10. Computers, monitors, imaging equipment, fax machines and other office equipment.	Note if IT power management is being used. Are the equipment energy friendly.

be largely dependent on how well the walkthrough and energy team performs and how well management provides leadership and support.

Walkthrough is the primary thrust of the ECW component. Walkthroughs specifics are:

Purpose: To identify energy waste and determine the appropriate fix.

Who? Facilities, engineering, technicians, energy team leader and others that can contribute.

What? Kickoff meeting (optional). Walk the facility and yard and record anything that uses energy—what it is, the amount of energy used (if possible), whether it can reach a state of excessive energy, what preventative maintenance is being performed now, and other pertinent information.

Also, in this phase energy awareness is developed jointly by the energy team and walkthrough team, although there could be several members on both teams. Walkthroughs should be done every 2-3 years. Instead of doing energy waste walkthroughs some organizations prefer to have an energy audit to identify the energy waste in their organization and recommendations to eliminate or minimize it.

PERFORMING AN ENERGY AUDIT

Buildings use nearly one third of the energy used in the United States every year. A typical energy audit cost is 5 percent of the organization's annual energy bill (Oppenheim, 2000). A good energy audit can eventually save an organization 15-20% of its annual energy costs.

An energy audit is a thorough and detailed examination of a facility's energy uses, consumption and costs, and its purpose is to identify recommendations to reduce those uses, consumption, and costs by implementing proposals of equipment upgrades, conservation practices, and operational changes. Energy audits identify cost effective opportunities that will probably result in significantly lowered electricity and natural gas costs (also steam, water and sewer costs if included in the energy audit scope).

There are three types of energy audits:
1. Walkthrough
2. Standard
3. Simulation

The walkthrough type has already been explained. The standard type explanation will follow. The energy audit process is:

Step 1. Preliminary coordination
management goal, purpose, type of audit, level of effort including, estimated man hours and total cost
Step 2. Data gathering and analysis
Facility & equipment & systems info, electricity, natural gas and other consumption data, maintenance programs

Step 3. On-site inspection
          Observing, checking equipment, systems, windows,
          doors, etc.
Step 4. Data analysis and evaluation of findings
Step 5. Write report and present to facility owner

An energy audit evaluates the energy efficiency of all building components, equipment and systems that impact energy use. The audit process should begin at the utility meters (electricity, natural gas, steam, etc. that are included in the audit scope) where the sources of energy come into the facility are measured and then show up on the appropriate energy bills. The energy usage of each type of energy is measured and analyzed to determine if any of the charges can be reduced. An assessment of the facility, systems, and equipment condition is made. The report will include recommendations on how to increase energy efficiency through improvements in preventative maintenance to include energy-centered maintenance, installing of energy-saving technologies and energy conservation measures (ECM) or energy projects with good payback periods.

The traditional energy report looks like this.

• Executive summary
    — Building/facility information
    — Utility summary
    — Energy conservation measures
    — Operation and maintenance measures
    — Appendices

By: B. Younger in 2000.

The author recommends using the table of contents shown below.

• Executive Summary
    — Building or facility information
    — Energy use, consumption and cost review
    — Known Energy Management Best Practices Review
    — Operations and maintenance
    — ECPs or ECMs recommended
    — Appendices

The report's major areas with proposed contents are shown below.

• Executive summary

— Purpose of the audit
— Type of audit
— Summary of findings and recommendations

- Building or facility information
  — Facilities' background
  — Description of facility or facilities
  — Operating schedule including occupancy patterns, number of people, number of computers and other relevant data

Energy Use, Consumption, and Cost Review

- Gather 2-3 years of utility bills for each energy used.

- Perform an energy review, establish an energy profile, and baselines.

- Do a pie chart of energy types used and their percentages.

- Do a pie chart of electricity consumption for HVAC, lights, office equipment, and other.

Operations and Maintenance

- Present O&M program

- Identify low- or no-cost opportunities to reduce energy.

- Check if an energy procurement policy has been designed and implemented.

- Determine if IT power management has been implemented.

- Determine if reducing office paper use has been accomplished.

- Has an energy conservation program been developed and implemented? Is it conclusive?

ECPs or ECMs Recommended

- Energy conservation measures that are potential energy projects with reasonable payback periods are recommended.

Appendices

Put in copies of any analysis, observations or charts that help explain recommendations included in the report.

An energy audit done by a professional, who is in this business, can save any organization by obtaining lower bills with less consumption.

ENERGY AWARENESS TRAINING

Energy awareness training should be provided annually. There-fore, it should be updated when something significant, like a success story, occurs.

All organization's personnel should receive energy awareness training either by computer or in a conference or classroom led by the energy champion or a member of the energy team.

Example:

POWER OF ONE

Often we have heard if you change your attitude, you will change your behavior. Changing our behavior is how we achieve a cultural change. We have seen this occur in recycling. Once all of us were briefed on the recycling program, what could be recycled, where the containers were located, and why we should recycle, most of our employees and contractors joined in the recycling effort. It is not unusual to see someone place a recyclable item in the trash and be corrected by another person to place it in the recycling bin or container. Why did this happen? Because it is simply the right thing to do. One ton of paper recycled saves 17 trees, 4,000 kWh of electricity, and 7,000 gallons of water. How about electricity? Have we reached a point where we turn off the lights when we leave the conference room or at night when we leave the office? Do we unplug our appliances such as radios, and other personal items when we go home at night? Do we ensure that our computers and monitors go to sleep when they are not in use after a short time? Are we thinking "let's save energy" and providing suggestions on how to do so to the building manager when we observe usage that is not necessary or can be reduced? Yes, one person can make a difference, YOU ARE A POWER OF ONE, and together we can make a significant reduction in energy consumption. Let our daily behavior be one of saving electricity. Once electricity is used, it is gone forever. Resources have been used and are not renewable. Save electricity because it is simply the right thing to do and, yes, you are a power of one.

Other contents should include:
- — Organization's energy reduction goal and renewal energy goal.
- — Why the organization has selected energy reduction as a corporate objective.
- — Energy policy.
- — Energy plan portions.
- — Energy performance indicators.
- — Energy reduction deployment process, the components and strategies.
- — The energy champion and the energy team members and how they can be reached.
- — Management's expectations of everyone's support and involvement.
- — Present success stories.

Chapter 4

Energy Centered Objectives (ECO)

SOME WORDS THAT CAUSE CONFUSION

These five words can be confusing when discussing objectives. They should be understood up front. They are defined below:

- Purpose—The reason you do something or plan to do something.
- Aim—What you hope to achieve by doing something.
- Goal—Something important that you hope to achieve in the future. Long-range or brief.
- Target—The exact result that you desire from doing something. What you are going to address and do to achieve an objective.
- Objective—A specific outcome you desire to attain or achieve. Described in an objective statement that tells what you want to happen.

The energy team will develop at least three to seven objectives the first year and at least three each year until the energy plan has been fully achieved. Typical objectives the first year are:

1. Determine the feasibility and benefits of implementing ISO 50001 Energy Management System (EnMS).
2. Develop an energy communications plan.
3. Design and implement a documentation system.
4. Determine where occupancy sensors should be installed, the number and cost.
5. Develop and communicate an energy conservation program.
6. Develop an internal audit of energy program.
7. Select performance measures, graph and communicate progress and results.
8. Determine the legal and other requirements.

These are just a few possibilities. Objectives can come from many areas. It is important for the energy team to establish objectives that move the energy management program forward. Objectives and targets must be established but not too many that will be very difficult to complete and may result in overwhelming the team members.

WHERE DO OBJECTIVES COME FROM?

Objectives and Target Ideas Come from Many Sources

Objectives come from many different sources. If it is something the energy team wants to accomplish to achieve a desired outcome, then it is an objective regardless of its source. Objectives can come from:

1. SWOT analysis.

2. Walkthroughs.

3. Implementing the low hanging fruit ideas.

4. Developing communications plans, monitoring and measuring plan, procurement plans, and others.

5. Implementing ECM—measures.

6. Management and employees' suggestions.

7. Developing and instructing training.

8. Developing energy performance indicators to include load shedding.

9. Energy functional teams.

10. Management/employee brainstorming sessions.

11. Key results areas (KRAs).

A Strengths, Weaknesses,
Opportunities and Threats (SWOT) Analysis

Let's look at the completed energy SWOT.

The Table 4-1 SWOT features strengths versus opportunities. Several possible objectives come to mind immediately from the SO portion of the SWOT analysis. For example some possible objectives are:

Table 4-1. SWOT

Strengths	Opportunities	Helps?	Other
1. Energy Reduction saves money.	A chance to save money that can be used for other mission requirements.	Yes. Can use this to help sell top management on reducing energy.	Includes a Force Field Analysis, derived from results of SWOT, which can help sell management.
2. Facilities personnel have some energy training and experience.	Gives organization a chance to contribute to the environment by saving energy.	Yes. Includes facilities personnel on energy team and walk through team.	Energy Team will need other cross-functional members.
3. Electric Utility does energy audits and assists in funding energy projects with excellent payback.	Can get assistance from electric utility and upgrade some of the old air conditioning and heating equipment.	Yes. Later in energy reduction deployment, ECP projects can be identified scoped, and payback period calculated.	After the low hanging fruit are picked and other plans developed, ECP will be addressed.

Strengths	Weaknesses	Helps?	Other
Reducing energy consumption and cost lends itself well to be a corporate objective that deployment can involve everyone in the organization. 2. ISO has completed a standard called ISO 50001 Energy Management System (EnMS) that can serve as a guide or actually be implemented if management desires.	There has been no past improvement effort of this magnitude attempted by the organization before. 2. Strategic planning has not been used by the organization on a regular basis.	Yes. Even though the organization is inexperienced in this area, this is an excellent initiative to gain the experience and be successful in saving money.	Possibly an external energy facilitator would be helpful and worthwhile.

Strengths	Threats	Helps?	Other
1. Facilities personnel have some energy management training and experience.	1. Insufficient supply of electricity available during high energy use days.	Yes. Facilities personnel can help plan the backup supply needed for critical activities.	Energy Team with facilities and engineering help should develop an emergency plan that will solve this problem.
2. Energy Reduction saves money.	2. Electricity and/or natural gas cost could significantly increase due to unplanned situations such as military conflict, storms, etc.	Yes. Means more savings provided the energy supply is available for the organization's demand.	Not very probable to happen, but has in the past.

Table 4-2. Possible Objectives from SWOT Analysis

Objective	Target	Remarks
1. Develop an Energy Reduction Plan.	10% Electricity Intensity reduction by end of 2016.	Target is SMART.
2. Put in place a cross functional energy team.	By Feb. 2015. Covers all functional areas.	Includes at least one member from Facilities or Engineering.
3. Determine the feasibility of using an UESC to fund and accomplish our energy projects.	Accomplish an energy audit and identify energy policies with payback of 3 years or less by Dec 2015	UESC is Utility Energy Services Projects
4. Design and develop energy awareness training.	Sent to all organizations management, supervision, employees and contractors by Dec 31, 2014.	Can be sent as an attachment to an email, a CD, or given live to groups until all have received the training.
5. Develop an energy emergency and contingency plan.	Include sufficient backup for all critical power needs by Dec. 31, 2015.	Will include different measures such as emergency generators and separate distribution feeder lines.

Walkthroughs

During the walkthroughs, numerous possible objectives and targets could be discovered. Below are just a few possibilities.

Table 4-3. Possible Objectives from Walkthroughs

Objectives	Targets	Remarks
6. Develop an energy conservation program and communicate to all personnel.	List ideas during facility walkthroughs and do Management/Employee Brainstorming Sessions, Establish Functional energy Teams, encourage suggestions and complete by April 2015.	Can be sent as an attachment to an email, a CD, or given live to groups until all have received the training. May also, put in an organization's procedure.
7. Put variable speed drives on all motors that can use them effectively.	All motors that do not need to run all the time.	Will need to include both an identification and installation portions.
8. Determine feasibility or a re-commissioning project.	Will need to develop scope and finish by June 30, 2015. Focus on Cost, Other Benefits, and Expected Savings.	Will be part of ECP.

Implementing Low Hanging Fruit Ideas

Over 80 possible objectives could be written alone for this area. All of them would be good, saving energy consumption and costs. Some examples are:

Table 4-4. Possible Objectives from Low Hanging Fruit

Objectives	Targets	Remarks
9. Install Occupancy Sensors in our Facility.	Hallways, Break Rooms, Rest Rooms, Mechanical Rooms, Copiers Rooms and complete by Dec 31, 2015	Calculate a payback period to help sell to management.
10. Implement a "Turn off the Light Program".	All the facility and by March 31, 2015.	Put up posters and place small stickers on the light switches.
11. Implement FAME at our facility.	By March 31, 2015. Excess electronics and appliances in work space areas.	Facilities and Management Evaluation

Developing Communications Plans, Monitoring and Measuring Plan, Procurement Plans, and Others

If your organization decides to implement ISO 50001 Energy Management System (EnMS), these objectives and targets will be mandatory to accomplish. They are:

Table 4-5. Possible Objectives from ISO 50001 EnMS

12. Implement a Communications Plan.	Include both internal and external communications. Complete by February 28, 2015.	Put into a procedure also.
13. Implement Correction Actions Plan.	Develop a Corrective Action Report and the Process to originate, complete and verify completion. By March 31, 2015.	Put in a Procedure also.
14. Implement a Measurement and Monitoring Plan.	Include Energy Measure whose development could also be an objective. Put any objectives when completed that need further tracking and monitoring. Put any reoccurring calibration requirements. Complete by April 30, 2015.	Put in a procedure also.
15. Implement an Energy Friendly Procurement Plan.	Include all electronics and compliances. Encourage Energy Star and EPEAT.	Put in a procedure and show as a policy.
16. Develop and implement a Documentation Plan.	All documents and records needed for an audit. Use organization's central management system or SharePoint. Complete by January 31, 2015.	Put into a procedure. Make readily available for others to use.

Implementing ECM—Measures

What are ECMs? An energy conservation measure (ECM) is any type of project achieved or technology implemented to reduce the energy consumption in a building or facility. ECMs can be in a variety of forms, water, electricity and gas being the main three.

The objective of an ECM should be to achieve a saving by reducing the amount of energy consumed by a particular process, technology or facility. The ECM has a calculated payback period, project description and scope, location and estimated cost, and it is kept in a government data system so if funds are available, they can be allocated to the most deserving projects.

Table 4-6. Possible Objectives from ECM

Objectives	Targets	Remarks
17. Calculate the payback period for the energy projects identified and put them into an ECM form.	Costs divided by savings for each project.	Either payback period or ROI (Return on Investment) are normally used to show energy projects attractiveness for funding.
18. Enter all the ECMs into the designated data base.	All identified organization's energy projects	Each energy project has a separate ECM and comes one line item in the data base.

Management and Employee Suggestions

Management and employees are encouraged to offer suggestions on energy to the energy champion or energy team members.

Developing and Instructing Training

Training is vital to any improvement initiative, and that includes energy reduction. Energy awareness training and energy conservation training are a must. However, there are other processes or specialty training that may be necessary.

Developing Energy Performance Indicators to
Include Load Shedding

Having energy performance indicators to measure progress and results is a "must" for energy reduction. Also, key players must understand the performance indicators.

Table 4-7. Possible Objectives from Suggestions

Objectives	Targets	Remarks
19. Add in the energy awareness training the encouragement of management, employees and contractors to offer energy suggestions.	Add the reason why suggestions are important: to meet our energy goals, to save money to be used in other mission areas, to save the environment, and to continuously improve.	Make the request compelling.
20. Develop a recognition program for energy suggestions from management, employees and contractors.	Make it Timely, sincere and appropriate for the benefits derived from the suggestion.	Does not have to be money as long as the three criteria in the targets are met.

Table 4-8. Possible Objectives from Training

Objectives	Targets	Remarks
21. Develop and present "How to Read an Electric Bill" to the accounting and facilities personnel.	By March 31, 2015. Explain all the adjustments and charges.	Include others that would like to receive the training
22. Develop instructions for calibrating the electric and natural gas meters.	All electric meters and natural gas meters. By April 30, 2014.	Meters need calibration periodically.

Table 4-9. Possible Objectives from Load Shedding and EnPIs

Objectives	Targets	Remarks
23. Develop data collection plans for each key energy performance indicator (EnPIs).	All EnPIs by Jan. 31, 2015. Include source of data, formula, frequency, who is responsible for collecting data, and what is or is not included in the data.	Makes having current performance indicators graphed easier and accurate.
24. Identify the loads to be shed into the appropriate categories.	Critical, Important, and Non-Critical Categories. Do by May 30, 2015	Critical loads are immediately powered after an interruption by emergency generators.

Energy Functional Teams

Functional teams are excellent to identify specific actions that can be done to reduce energy consumption in their work area.

Table 4-10. Possible Objectives from Energy Functional Teams

Objectives	Targets	Remarks
25. Increase the thermostats by 2 degrees F in the summer and decrease in the winter by same amount.	68 degrees F in winter and 78 degrees F in summer. Accomplish by Oct 15, 2015.	Employees may have to dress different.
26. Develop instructions of what appliances and electronics can be included in the work station and put into an organizational procedure.	List only those authorize to have in work center and include those that are not such as refrigerator, space heater and electric fans.	Put into a procedure.

Management/Employees Brainstorming Session

These brainstorming sessions can create a lot of good energy savings ideas. In addition, they get employees involved. Once involved, employees support the energy reduction initiative.

Table 4-11. Possible Objectives from Brainstorming Sessions

Objectives	Targets	Remarks
27. Develop a management/employee brainstorming sessions program.	Agenda. What functional areas. Time schedule. Managers for sessions and facilitators. Book the conference rooms. Write introductory letter. Arrange for pads on easels and pens/markers. Complete by Feb. 15, 2015	Give energy awareness training first. Next, write on a pad or white board "How can we reduce Energy Consumption in your work areas?"
28. Determine the feasibility of using lap top computers instead of desktop ones.	Do a cost comparison and an energy savings analysis. Determine how program would best be phased in if found feasible. Accomplish by June 15, 2015.	Lap tops use less electricity than desk tops.

Implementing ECM and ECM—Measures

Implementing energy centered maintenance and/or energy centered measures could necessitate establishing several objectives and targets.

Key Result Areas

Key results areas (KRAs) are the key areas of an organization, related to the mission, that are best to target for improvement. The ten most common KRAs are:

- Quality
- Sales
- Cost
- Services

Table 4-12. Possible Objectives from ECM & ECM—Measures

Objectives	Targets	Remarks
29. Perform the walkthrough to identify what should be included in the ECM.	Identify items that do not fail but reach an excessive use of energy. Complete by: April 30, 2015.	ECM is Energy Centered Maintenance and follows the RCM (Reliability Centered Maintenance)
30. Determine the preventative action for each item to be included in ECM.	Specific what should be done, frequency, craft, labor and materials and cost. Complete by May 30, 2015	Ensure the action is cost effective.
31. Develop the maintenance checklists for ECM.	Do by area. Ensure no more than 80% of a craft's time is on a particular checklist. Complete by July 31, 2015.	Checklists will help develop the schedules.
32. Develop Short Interval Schedules for ECM.	Accomplish by craft a week's worth of work in accordance with the ECM maintenance plan.	Can be for one shift instead of a week.
33. Select the CMMS for our organization.	Select one that interfaces with other organization's systems and meets your specifications. Accomplish by August 31, 2015.	Organization may already have a system that can be used.
34. Implement IT Power management program.	Monitors set Energy Star "Sleep Function" and "Hibernate" for CPUs. Complete by June 30, 2015	Monitors after 15 minutes. CPUs after being idle for 30 minutes.
35. Implement Reduce Office Paper Initiative.	Identify ways to include duplexing and use of electronic files. Implement them.	Calculate cost savings and environmental savings.

- Delivery
- Schedule
- Timeliness
- Stewardship
- Safety
- Customer Satisfaction

Let's add energy as a KRA and determine its relationship to several of these important KRAs.

It should be obvious that objectives can come from numerous areas. Once the objective is established, then the target is developed to show what is going to be done to meet the objective.

Figure 4-1. Energy & Other KRAs

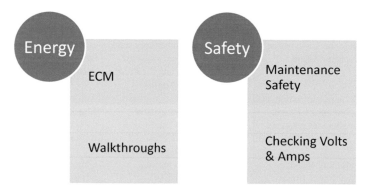

Figure 4-2. Energy and Safety Relationship

Figure 4-3. Energy and Quality Relationship

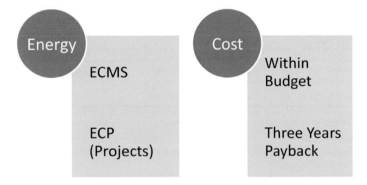

Figure 4-4. Energy and Cost Relationship

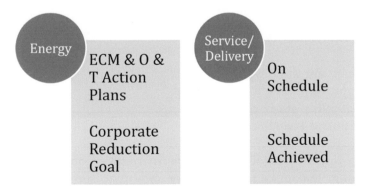

Figure 4-5. Energy and Service/Delivery Relationship

Table 4-13. Possible Objectives from KRAs

Objectives	Targets	Remarks
36. Achieve Excellent Safety Results in Performing Maintenance (ECM and PM) and Voltage & Amps Readings During Walkthroughs.	Safety Principles and Plans for ECM & PM and Voltage & Amps Readings whenever they occur.	Safety first
37. Meet or Exceed Stakeholders Expectations.	Management, Employees, On Site Contractors, Owners, Clients	A customer satisfaction survey can be administered periodically.
38. Verify all Energy Results to Ensure They are Real.	Use Verification and Measurement Tools & Techniques on Energy Results	Heat and Cooling Days compared to Baseline. Production Compared to Baseline.
39. Each Year, Ensure all Energy Costs are Within Budget.	Focus on Budgeted Large Items and perform cost control.	Stay on or below Budget
40. For each Project, Calculate the Simple Payback Period	Each Project Considered for Funding and Contract.	Can Bundle Projects
41. Ensure ECM is Accomplished On Schedule.	Develop Schedule and Monitor Progress and Take Action if needed.	Use project management techniques.
42. Ensure the Energy Reduction Goal is Attained on Schedule.	Develop Schedule and Monitor Progress and Take Action if needed.	Use project management techniques.

It should be obvious that objectives can come from numerous areas. Once the objective is established, then the target is developed to show what is going to be done to meet the objective.

TYPES OF OBJECTIVES

There are three types of objectives. They are:
1.   Improve or develop something
2.   Feasibility study or research
3.   Maintain something

To improve objectives can be to reduce something, increase something, maximize something, or minimize something. Using an objective template, let's see what an improvement type objective will look like.

Figure 4-6. Improvement

Objective _____

Facility name: Thomas Building   Document #: OT-2015-001

Objective: To reduce energy consumption in the bldg. by 5% by end of 2015

Target: Install occupancy sensors in hallways, restrooms, break rooms, mechanical rooms, and copier areas

Initiation date: January 15, 2015

Anticipated completion date: December 31, 2016

Actual completion date: _____

Electricity high users: Electric lighting

Baseline: Calendar year 2014   Monitored or measured: yes/EnPI

Table 4-14. Action Plan for an Improvement Objective

Required Action	Responsible Person	Target Date	Status (Red, Yellow, Green)	Comments
1. Identify Areas where OCs will go.	Facility Specialist	By Feb 15, 2015		
2. Determine Number and Estimate Energy Savings	Facility Specialist	By March 10, 2015		
3. Research Price and Present to Mgt for Approval	Facility Specialist	By April 8, 2015		
4. Contract for OC and Installation	Procurement	By May 10, 2015		
5. Install & Verify	Contractor/Fac. Spec.	By Aug. 10, 2015		

Figure 4-7. Feasibility or Research Objective

Objective and Target

Facility Name: Thomas Building   Document #: OT-2015-002

Objective: To determine which outside company to hire for the energy audit.

Target: Hire a company with professional energy consultants with electrical and mechanical engineering expertise. Meet our requirements plus identify savings at least 10 times greater than their fee.

Initiation Date: February15, 2015

Anticipated Completion date: August 31, 2015

Actual Completion Date: _____

Electricity High Users: HVAC, Electric Lighting, Production, Office Equipment

Baseline: Calendar Year 2014   Monitored or Measured: Yes/EnPI

Table 4-15. Action Plan for a Feasibility or Research Objective

Required Action	Responsible Person	Target Date	Status (Red, Yellow, Green)	Comments
1. Identify Companies Who Perform Energy Audits in Surrounding Area.	Energy Team Leader	By March 1, 2015		
2. Develop Criteria For Meeting Our Requirements	Energy Team Leader with Coordination with Energy Champion	By March 15, 2014		
3. Call and Determine Which Ones Meet Our criteria or Specs	Energy Team Leader	By March 20, 2015		
4. Select Company to Do Energy Audit	Energy Team & Procurement	By May 10, 2015		

Table 4-16. Action Plan for a Maintain Objective

Required Action	Responsible Person	Target Date	Status (Red, Yellow, Green)	Comments
1. Read the Emergency Plan and Identify Areas that Need Updating	Energy Champion	By Feb 3, 2015		
2. Gather the new data and info to upgrade the plan.	Energy Champion	By March 5, 2015		
3. Make the revisions	Energy Champion	By April 30, 2015		
4. Publish plan and brief the impacted on the changes	Energy Champion	By June 1, 2015		

As shown in Table 4-16, objectives can originate from many areas or sources. Every objective is one of three types which was demonstrated above. A simple checklist to help in preparing an objective and target and action plan is shown later in Figure 4-11.

Figure 4-8. Maintain Objective

Objective and Target
Facility Name: Thomas Building   Document #: OT-2015-003
Objective: To maintain the organization's emergency plan.
Target: Correct, accurate, and useful emergency plan.
Initiation Date: January 22, 2015
Anticipated completion date: December 31, 2016
Actual Completion Date: _____
Electricity high users: N/A
Baseline: N/A   Monitored or Measured:
                                                 Yes/Complete or not

Figure 4-9. Objectives and Targets Checklist

Objectives and Targets Checklist—Yes or No
1. Is the objective an improve/reduce, feasibility study or maintain objective?
2. Does the target specify what is going to be addressed and how much increase or decrease and by when?
3. Does the O&T contribute to meeting the goal.
4   Do the activities tell <u>what</u> is going to be done by <u>whom</u> and <u>when</u> it is expected to be completed?
5. When the actions or activities are complete, is the target achieved?

ENERGY OBJECTIVES AND TARGETS AND MAKING TARGETS OR OBJECTIVES SMART

The objective can be simply one of the 42 possible objectives listed above. The target identifies what is going to be addressed to accomplish the objective. The targets should be made SMART—Specific, Measurable, Actionable, Relevant and Realistic and have a Timeframe. Reduce energy consumption by 5% by developing and implementing an energy conservation program by Dec. 31, 2014. The action plan shows what is going to be done, by whom, and when. Often the status using the green light approach is also shown on the action plan. (Green—On schedule, Yellow—A little behind schedule, and Red—A lot behind schedule).

Figure 4-10. Objective and Target for Legal Requirements

Objective: Determine the legal and other requirements for an energy reduction program.

Target: Identify any federal, state, county, city, or local legal requirements by Feb. 20, 2015.

Objective number: OT-14-01 (First OT for objective and target—the last two digits of the year the O&T is originated—the number of the O&Ts as they are originated—the first one starts with 1, then 2 and in sequence as they are originated)

Responsible person: Jerry Johnson

Date expected to be completed: Feb. 20, 2015

Measure of luccess: A list of legal requirements is generated or not.

The status should be written in the O&T template at each energy team meeting. Any barriers, or unusual events should be placed in the remarks if it contributes to understanding the status of the O&T.

Checks:

1.   If the actions in the action plan are done will that complete or satisfy the target?
2.   If the target is met, will the objective be obtained?
3.   If the measure is achieved, will the target be achieved? If not, make the target SMART—specific, measurable, actionable, realistic, and time framed.
4.   Are the actions achievable and within the expected timeframe?

If any answer for 1, 2 and 4 is no, then review and change the O&T template so that it can be answered yes.

In ECO and ECP is where energy efficiency is planned and implemented. Energy efficiency is using less energy to provide the same service or to achieve the same mission. Efficient energy use is called achieving energy efficiency. It can be defined as the goal to produce products and services using less energy.

DON'T CHASE TOO MANY RABBITS

Introduction

In the identification of objectives and targets above, the author could identify hundreds of possible objectives and targets. It is easy for an energy team to feel overwhelmed with all the opportunities, issues

Table 4-17. Actions Plan Legal Requirements

What	Responsible Person	Completed by When?	Status-Green-Yellow-Red	Remarks
1. Research Federal requirement for energy and document applicable requirements	Jerry Johnson	Set 15, 2014		
2. Research State legal requirements for energy and document	Jerry Johnson	Oct 20, 2014		
applicable requirements.				
3. Research County legal requirements for energy and document applicable requirements.	Jerry Johnson	Nov 12, 2014		
4. Research city and local legal requirements and document applicable ones	Jerry Johnson	Dec 20, 2014		
5. Research organization for any applicable requirements and document.	Jerry Johnson	Jan 11, 2015		
6. Put on one list the applicable legal and other requirements by category(federal, state, county, city, local, and other)	Jerry Johnson	Feb. 20, 2015		

and things that need to be done to reduce energy consumption at their location. There is so much to research—so many things to learn, people to train, so much data and so many suggestions. People need to be kept up-to-date, and sometimes teams feel pressured to obtain early results. The energy team members are normally only part-time on a team. They have regular jobs to also worry about. The energy champion can be either part-time or full-time, depending on the size of the organization. It is not unlikely to find some energy teams trying to do everything and ending up accomplishing nothing or almost nothing.

Chasing Too Many Rabbits—Focus on the Fat Rabbit

Professor Noriaki Kano, a JUSE (Japanese Union Scientist and Engineers) counselor, in a training session preparing FPL for applying for the Deming prize in the 1980s, advised the attendees (over 300 including the author) not to chase too many rabbits. His example was if you went into a room filled with rabbits and were given the task of catching them, taking them outside, and placing them into a large container, what would happen? You would probably starting chasing rabbits near you and they would start running and jumping and you would not catch one single "cotton picking" rabbit. This could go on until you were exhausted and feeling like a failure at this assignment. "You are chasing too many rabbits." How do you solve this problem? Look in the room and locate the fattest rabbit in the room. Ignore the other rabbits and focus on the fat rabbit. Catch the rabbit and take him through the door (without letting any of the other rabbits out). Now come back into the room. What are you going to do now? Yes, you are going to locate the next fattest rabbit and focus on catching it. You do this until all the rabbits are placed in the outside container.

What did you do? You developed a strategy, turned it into an action plan, and then implemented or executed the plan. You focused on catching the fat rabbit. This simple story can help many teams, energy or otherwise, to get on track with a clear focus and produce desired results.

Professor Kano also called the first bar on a Pareto chart of opportunities or problems the Fat Rabbit. It is the one that will give you the most results. Once you eliminate it or reduce it below others, you then take the highest one left and work on it. That is now the fat rabbit.

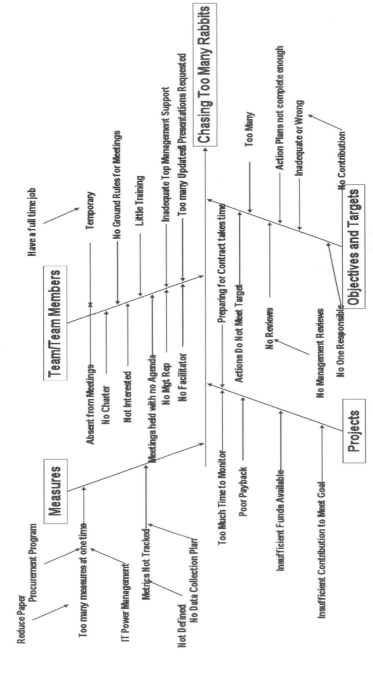

Figure 4-11. Fishbone Diagram

Focusing on the Fat Rabbit

When an energy team realizes they are not being productive, what should they do? The steps are simple but very important.

Step 1. Figure out the problems or causes.

Step 2. Fix them.

Step 3. Develop a strategy and plan and then implement it to get back on track to an efficient and effective team.

Step 1. Figure out the problems or causes

The best tool to identify root causes is the cause and effect diagram called the fishbone. (See Figure 4-11.)

Step 2. Continue finding the possible causes and then fix them

A root cause matrix is an excellent tool to accomplish Step 2. Identify the most probable causes and start evaluating the root cause matrix. Often it is good to start with categories such as team leader/team members and address possible root causes of problems in this area.

Table 4-18. Team/Team Members Root Causes Matrix or Table

Team Members Causes	Countermeasures	Team Meeting Causes	Countermeasures
Part Time Assignment	Discuss with Member & His/her Boss to Gain Commitment	No Agenda	Use PAL for all meetings
Little Training	Provide Training-ECMS, O & Ts, Meetings Effectiveness	No Facilitator	TL(Team Leader) or appoint one
Absenteeism	Keep Track, Discuss, Put in meeting minutes	No Management Representative/Energy Champion	Team Leader discuss with management
Some not interested	Make meetings interesting & productive-Establish roles and responsibilities	Meetings are non-productive but still have many of them.	Use PAL and plan meetings better. Have a trained facilitator.

To keep these problems from occurring, management should implement the following actions at the start of the energy reduction effort. Management needs to appoint a management representative for energy. He or she is a person in management that is a leader and interested in energy management. Their main duties are:

- Develop a charter for an energy team.
- Formulate a cross-functional energy team with a facilitator and at least one member from facilities/engineering.
- Keep top management informed on issues, problems, progress and successes.

The management representative needs to form a cross functional energy team.
- Include 5 to 10 members with at least one member from facilities or engineering.
- Conduct team building to get team working well together.
- Form "Storm, Norm, Perform During Storm, and Norm" develop ways to resolve conflict, communicate effectively, and become goal driven and a high performance team.

The energy team should:
- Hold meetings at least monthly.
- Prepare an agenda using PAL (purpose, agenda, limited).
- Develop and use EnPIs.
- Keep minutes using PAPA (purpose, agenda, points, action items).
- Stay focused and overcome any barriers until goal has been achieved.

Provide the team with training:
- Energy awareness
- Energy conservation
- Objectives and targets and actions plans development
- Effective meetings management

The energy team should establish no more objectives and targets than they have team members. Assign the objectives to different members to be the person of responsibility. Otherwise, the team could quickly be chasing too many rabbits.

In developing objectives and targets, use the below checklist:

1. Is the objective an improve/reduce, feasibility study or maintain objective?

2. Does the target specify what is going to be addressed and how much increase or decrease and by when?

3. Does the O&T contribute to meeting the goal.

Table 4-19. Root Cause Matrix for Objectives and Targets

Objectives and Targets	Countermeasures	Actions Plans	Countermeasures
Too many.	Limit to no more than one per team leader plus team members	Not Complete	Get Training
Wrong ones. Not meaningful.	Develop a checklist and use	Do not meet targets	Use Checklist
No one responsible.	Assign one member of the team to each O &T. A member can have two O & Ts if appropriate.	Not detailed enough to adequately show what is needed to be done.	Use Gantt Chart if necessary.
No Reviews.	Do Team reviews & conduct Management Reviews annually.		

4.   Do the activities tell **what** is going to be done by **whom** and **when** it is expected to be completed?

5.   When the actions or activities are complete, is the target achieved?

If the answer is no on any of the questions, redo the objective. The last category is measures.

The countermeasures for measures root causes are explained on the matrix above.

Chasing too many rabbits causes lost productivity, frustration, chaos, and poor quality. Implement the countermeasures to eliminate the root causes or at least minimize them. Remember focus on the fat rabbit—one at a time.

Having excellent objectives and targets increase energy performance while proving momentum for the energy team and the deployment effort.

Table 4-20. Root Causes for Projects

Project Causes	Countermeasures
Takes a lot of time to prepare contract	Get an engineer or facilities person on team and let them coordinate with design and others as necessary.
Scarce Funds	Calculate the payback period and try to get several projects under three years or less.
In sufficient Contributions to meet corporate target on time.	Add up Contributions from ECO, ECM-Maintenance, ECM-Measures and subtract from Goal. This is how much the projects (ECP) will need to cover.

Table 4-21. Root Causes for Measures

Measures Causes	Countermeasure
Trying to do everything at one time with no additional help.	Do one at a time. Get a Middle Manager to help on each. Guide them & develop action plan together & let them implement.
Metrics not tracked or graphed	Track & Graph Monthly KWH Consumption and CF for Nat Gas Consumption. Track & Graph Intensity. Do ELF Quarterly and Renewable % semiannually.
No data collection plans	Develop a Data Collection Plan for each indicator. Where data is located, who will collect it, what is the data to be graphed, the frequency and units of measure along with the formula.

Chapter 5

Energy Centered Projects (ECP)

INTRODUCTION

The energy centered projects (ECP) purpose is to identify projects that are justifiable and will reduce energy use. Figure 5-1 shows the elements for this component and the activities that need to be accomplished to place projects under contract. Once under contract, project monitoring and management to include verification of the product or service received must be accomplished.

The walkthrough team, energy team and facilities or engineering will identify projects that need to be accomplished to eliminate energy waste. For example, if the data center is cold, then a project to separate the cold air from the hot would be so beneficial that it could save 30% in data center electricity use. Reducing the temperature in the data center but staying in the standard and changing lights from T-12s to T-5s or using LED could save 5-10% or more of the electricity usage. These projects will need specifications, statements of works, advertising and contracts awarded, provided they are funded by top management. To be funded, management will need to see a projected payback period (how many years it will take for a project to pay for itself). Payback is savings or cost avoidances/total cost. A good payback period is 3 years. However, the government funds projects in reducing energy use

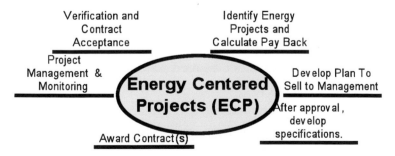

Figure 5-1. ECP Elements

up to 11 years. Utilities may fund projects with greater than 3 years' payback. Contacting an ESPCO (Energy Services Performance Contract Office) or a UESC (Utilities Energy Services Contract) to see if they will fund your projects could be a beneficial step. Projects with more than a 3-year payback can be bundled with those that do have a 3-year payback and still be accomplished.

Once the projects have been funded, project specifications will be needed. Specifications (often abbreviated as specs) refer to a specific set of requirements to be met by material, design, product, or service. Once the specs and then the design are ready and the funds available, a request for bids is sent to perspective bidders or placed on a website or in newspapers, so the contractors can prepare and send the organization or company proposals. The organization reviews the proposals and awards a contract, normally to the least bidder.

The contractor accomplishes the contract. The organization monitors progress and inspects the work to ensure conformance to the specifications. Project management is the monitoring, reviewing, inspecting, and evaluating that helps ensure the project is completed on time, within cost and of quality. If the contract meets these key elements and meets the organization's measurement and verification efforts, then the contractor is paid.

POSSIBLE PROJECTS

Some possible energy projects are:
1. Replace windows with double paned windows.
2. Perform an energy audit of the facilities.
3. Replace cooling tower.
4. Add ventilation to certain areas of the facility.
5. Add insulation to the walls and roof.
6. Replace roof with either a green roof or a white roof.
7. Install capacitors to eliminate the low power factor.
8. Change lights from T-12s to T-8s or T-5s and their ballasts.
9. Install ceiling fans in large open areas to reduce natural gas usage of the heaters used.
10. Install occupancy sensors throughout facility.

11. Replace security lighting.

12. Remove machines that are no longer required.

13. Upgrade boiler with new switches and sensors.

14. Upgrade building automation system.

15. Install new advanced meters.

These are just a few of the possible projects that a facility may need to accomplish. Projects will vary by facilities and age of the equipment and infrastructure. Renewable energy strategies will be accomplished here in accordance with the energy plan.

Each project should have the payback period calculated and used as the primary selling point to management for funding approval. Payback is a breakeven point. The payback period is the costs divided by the savings. For example, an energy project for lights reduced kWh usage by 181,000 kWh a year. The cost per kWh is 7.5 cents. The savings is .075 times 181,000 = $13,575.00. The cost of the lighting project was $30,000. The payback period is cost/savings = $30,000/$13,575 = 2.21. The lighting project is projected to pay for itself in 2.21 years. After 2.21 years, an annual savings of $13,575 will be realized. Normally, management will try to fund any energy projects with a payback of 3 years or less. In government, an executive order has allowed up to 11 years payback for an energy project.

In government, unfunded projects are first put on a basic project information form (an ECM, or energy conservation measure form) and then placed in a data base so they can be tracked and monitored.

ENERGY SERVICES PERFORMANCE CONTRACTS (ESPC)

An ESPC (energy services performance contract) is a contracting vehicle that allows agencies to accomplish energy projects for their facilities without upfront capital costs and without special congressional appropriations to pay for the improvements. Some utilities furnish the same services as an ESPC. They are called a UESCs. An ESPC project is a partnership between the customer and an energy services company (ESCO). The ESCO conducts a comprehensive energy audit and identifies improvements that will save energy at the facility. The

ESCO designs and constructs a project that meets an agency's needs and arranges financing to pay for it.

The ESPC process is shown below:

Phase 1. Project Planning—Interagency agreements and statements of work for services from FEMP.

Phase 2. Initial Project Development—Kick-off initial proposal/review, do RFP (request for proposal) template, risk & responsibility matrix.

Phase 3. Negotiation and award of final delivery order—detailed energy survey, final proposal/review.

Phase 4. Implementation and Performance Period—Commissioning & project acceptance.

Phase 5. Performance Period—Measurement and verification—pay if savings are materializing.

Some utilities provide these services. Check with your utility to see if funding and project support is available and if so, learn their participation process.

Chapter 6

Energy Centered Maintenance (ECM)

ECM ELEMENTS

The primary purpose of ECM is to reduce energy use by identifying equipment or items that can become energy hogs while still performing their function and then prevent that from occurring or stop it when it occurs. It is one of three methods that energy waste (ECW), once identified can be fixed—either eliminated or minimized. The other two "fix it" avenues are energy centered objectives (ECO) where an energy team develops objectives, targets and action plans, and where energy centered projects (ECP) have specifications developed and payback periods calculated. Worthy projects are funded and contracts let including recommissioning. These four areas, along with the planning and developing component, energy centered planning & development (ECPD), compose a simple but effective system, energy centered management system, for reducing energy use, consumption and cost.

A motor that runs for 24 hours instead of the required 7 hours because of a faulty switch is still meeting its functional requirement but is using excessive energy. The objective of ECM is to find these kinds of situations, if they are occurring, and perform preventative or

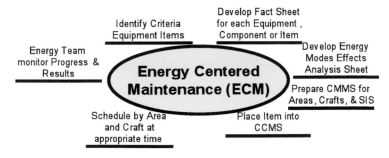

Figure 6-1. ECM Elements

corrective maintenance to keep them from becoming a reality. ECM is not a duplication of preventative maintenance but can be an additive activity or already included for another purpose that is to keep the equipment or component from failing.

WHY ECM?

Below are three primary reasons ECM is needed:

➢   Poor maintenance of energy-using systems, including significant energy users, is a major cause of energy waste in the federal government and the private sector.

➢   Energy losses from motors not turning off when they should, steam, water and air leaks, inoperable controls, and other losses from inadequate maintenance are large.

➢   Energy losses from conditions that can be altered such as hot air and cold air in a data center mixing and requiring excessive energy to cool the cold air that mixed with the hot air.

OBJECTIVE OF ECM

The objective of ECM is to ensure energy is not wasted thereby keeping energy costs from being excessive. A byproduct is that it can also reduce greenhouse gasses while reducing energy waste.

ECM supports the organization's preventative maintenance and/ or predictive maintenance program as well the reliability centered maintenance (RCM) program. Possibly some of the equipment or components designated to "run to failure" could be included in ECM if they have the potential to use excessive energy prior to failure.

If an organization is implementing ISO 50001 Energy Management System (EnMS), ECM can be incorporated into the planning phase with identifying the significant energy users (SEGs).

O&M SAVINGS

It has been found that O&M programs designed for energy efficiency can save 5% to 10% on energy bills without a significant capital investment and with just a little cash outflow. It is believed ECM,

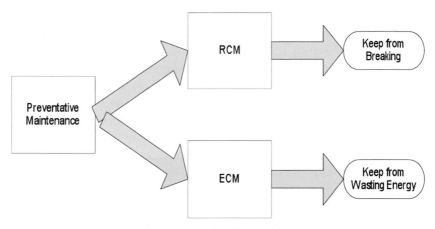

Figure 6-2. ECM vs. RCM

including an energy awareness and conservation program, can easily achieve these savings.

ECM AIM, COMPONENTS, AND DESIGN

ECM defined:
- ECM is not aimed at the probability equipment or component failure, but at using more energy than required to function.
- ECM is not a RCM project.
- While the preventive or corrective action may already be in the preventative maintenance program, no action is required. If otherwise, it is added to the program and work schedule.

ECM consists of four components shown in Figure 6-3.

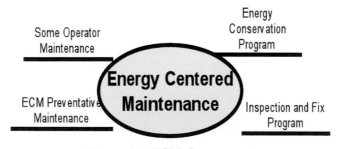

Figure 6-3. ECM Components.

The operator maintenance scope depends on the equipment complexity, cost or worth of equipment, and equipment needs such as fuel and cooling, operating hours, design, and other important characteristics. Operator maintenance is defined upfront when the equipment arrives and is installed. Operators are trained on what to do and how to do it. Most of the operating maintenance is to keep important fluid levels adequate and ensure equipment will run smoothly and not fail. ECM comes into play for only the operator maintenance that impacts or supports good energy use, for example, good gasoline and oil that help the engine and provide excellent operation time or mileage.

Operators should be trained in the ECM concept and objectives and look for ways to positively impact the program. Their ideas and suggestions should be sought.

Energy conservation is a basic unit of ECM. It is cost effective in that it does not include any high labor costs and materials. Most conservation actions are relatively inexpensive or would already be purchased if an energy conservation program were not implemented. Energy conservation includes focus on turning off the lights, unplugging electronics and appliances when not in use, using power chords that turn off when not in use, using daylighting when possible, planting trees and shrubs where they can block the sun, fixing fans to turn the most effective way, and setting the temperature to 78° in summer and 68° in winter. Also needed are monitors, duplex of office paper at the imaging machines, and implementing an Energy Star procurement program. These measures will be further explained later. In other words, doing what any organization's personnel can do on their own to save electricity consumption.

Inspection and fix programs that are scheduled by area or processes are an essential portion of saving energy. Replacing filters or cleaning lights are examples of maintenance that can be added to this function or kept in the preventative maintenance area. The inspection area can also be used to identify the organization's significant energy users and the equipment and components that belong in the ECM program. Leaks in windows, doors, or other areas can be detected and fixed.

Equipment, building components, and other items can deteriorate or be affected by a switch or sensor not working. The energy waste types encountered so far are listed in Figure 6-4.

Figure 6-4. Energy Waste Types

1. Running when it should be turned off
2. Having to work harder
3. Wear
4. Corrosion
5. Leaks
6. Stuck in the open or closed position
7. Not performing
8. Over performing
9. Overheating
10. Waste buildup
11. Incorrect settings
12. Switch or sensor failure (broke or seized)
13. Over lubricating

In the future as organizations implement, ECM this list should grow. In applying ECM later, examples will be shown for each of these energy wastes or excessive uses of energy. ECM includes routine inspections and then fixing any problems encountered. Signs to look for are:

- Noise
- Vibration
- Dirt
- Corrosion
- Running hot
- Sparking
- Deteriorating insulation
- Over lubrication
- Excessive friction and sparking
- Amps above manufacturer's recommendation

ECM activities for preventative or corrective actions are added to the organization's regular preventive maintenance, predictive maintenance or RCM programs. Often the maintenance activity to keep equipment from using excessive energy is the same fix to prolonging the equipment's useful life.

ECM is designed to keep equipment from using and wasting energy. It is not aimed directly to keep the unit from failing; however, indirectly it may. Example: a motor is running 24 hours and should run only a portion of that time. It is running because of a switch being stuck open. The motor has not failed but is wasting energy. Preventing this waste is the main thrust of ECM. However, it may also prolong the life of the equipment or component.

THE ECM PHASES

ECM is based on a concept similar to reliability centered maintenance. (RCM). RCM is focused on least cost to prevent equipment failure. ECM focus is on eliminating energy waste. ECM consists of seven phases.

Phase 1—Equipment, Function, & Energy Waste
Identify the Equipment or Component
Phase 1. Walk through the facility and identify the equipment. Then decide if components of the equipment are better to track and monitor.

For motors and equipment, measure energy efficiency and include in ECM if the energy waste is high and a preventative maintenance action can prevent excessive energy use.

Answer these questions and document:

Equipment or Item? Can this equipment or other item still operate and meet its function but can waste additional energy if not maintained?

Phase 1	Idendity the potential item
Phase 2	Test to determine any problem
Phase 3	If selected, determine what maintenance, repair or replacement is necessary
Phase 4	Ensure ECM is cost effective
Phase 5	Determine craft, frequency and labor required
Phase 6	Develop maintenance checklist and add to CMMS
Phase 7	Schedule work and monitor

Figure 6-5. ECM Seven Phases

Function? What is the item supposed to do?

Condition? In what condition is the item in that causes abnormal or excessive energy use?

Root Cause? What was the root cause that led to the item wasting electricity or other energy or water? The answer to this helps identify the countermeasure that is needed to put in place.

Energy Waste Cause? In what ways can it fail to a state that uses additional, wasteful energy but still function?

Preventatives? What systematic task can be performed proactively to prevent, or to diminish to a satisfactory degree, the consequences of the excessive energy waste?

Craft/Frequency/Labor Time? What craft is needed, the time to accomplish and what frequency should the work be accomplished?

Cost Effective? Is the cost of performing inspection and maintenance less cost than the potential to be saved?

Risk? What is the risk of not doing something?

Identifying Energy Waste:
Perform an Energy Efficiency Analysis of Selected Items Where Practical
Energy efficiency is defined as "To do the same mission or processes with less energy consumption."
Energy efficiency is calculated as
EE = (energy input – energy waste/energy input) × 100

The process to follow is an action plan including this process follows:
Step 1. Look at name plate—get energy information on energy input.
Step 2. Measure amps & volts and calculate Watts by multiplying the two. Watts = amps × volts
Step 3. Divide by 1000 to get kilowatts (kW)
Step 4. kW measured – kW name plate = kW of energy waste
Step 5. EE can now be calculated

Phase 2—Tests to Determine Problem
Phase 2 is shown in Figure 6-7, Tests to Determine Problem.

Required Action (What)	Person Responsible (Who)	Target Date (When)	Status	Comments
1. Get Energy Input Info From X-Ray Machine	Energy Manager	Jan 15, 2015		
2. Measure Amps & Volts	Electrician	Jan 15, 2015		
3. Calculate Watts & KW	Energy Manager	Jan 20, 2015		
4. Calculate KW of Energy Waste	Energy Manager	Jan 20, 2015		
5. Calculate Energy Efficiency	Energy Manager	Jan 20, 2015		

Figure 6-6. Energy Action Plan

Phase 1 Identify the potential item

Phase 2 Test to determine any problem

Figure 6-7. Tests

ECM—Locate and Inspect—Test for the Energy Waste—Determine Course of Action—Repair? Perform Maintenance? Replace? Let Run to Failure? Put into PM Program? Put into PM program but as an ECM Item? Power Quality?

If it meets the ECM test of using excessive energy, develop information for ECM and place in program. Develop a short fact sheet that explains what the function is, how the item can use excessive or waste energy and the cause that can make this happen.

Things to look for when inspecting:
* Noise
* Vibration
* Dirt
* Corrosion
* Running hot
* Sparking
* Deteriorating insulation

- Over lubrication
- Excessive friction
- Amps above manufacture's recommendation

Accomplish routine tests to determine energy waste or problem. They are:
- Observe and look and listen for signs of problems.
- Measure amps pulled and compare to nameplate.
- Measure volts and compare to nameplate—look for power quality issues.
- Test all key sensors and switches to ensure they work.

Other more technical tests are:
- Vibration—Install on all critical rotating machines.
- Infrared—Used to find hot spots (in electrical equipment such as transformers, switchgears and cables). Technology used to spot leaks in windows and doors.
- Megger testing—Detect grounds, damp windings, damaged insulation, current leakage to grounds.
- Thermo vision—Used to test the amount of heat or the heat flow through a piece of equipment and measure its quality for evaluation. Will pick up hot spots in electrical equipment such as switchboards, cables, etc.

Phase 3—Preventive Activity/Task Information
Determine the preventative or corrective task characteristics. Determine the preventive measure to prevent the failure or waste.

- Define the preventive task that when accomplished will either prevent the waste or identify the fix needed to do so.

Phase 1	Idendity the potential item
Phase 2	Test to determine any problem
Phase 3	If selected, determine what maintenance, repair or replacement is necessary

Figure 6-8. Phase 3 PM Activity

- The preventive maintenance tasks will be either a check or inspection and then a fix such as adjust, repair, or replace a component or sub-component.

- The preventive task should be aimed at the energy waste cause process, be very specific, and include specifications and tolerances where appropriate.

Phase 4—Estimate Possible Energy Savings & Determine Cost Effectiveness

Phase 1	Idendity the potential item
Phase 2	Test to determine any problem
Phase 3	If selected, determine what maintenance, repair or replacement is necessary
Phase 4	Ensure ECM is cost effective

Figure 6-9. Phase 4—Determine Cost Effectiveness

- Estimate the kWh that can be saved if the equipment goes into the energy waste cause mode. You may need to have the electricians measure the kW of the equipment first, unless there is a meter or an output like that on UPS. Multiply the kWh estimated to be wasted times the cost per kWh the organization pays for the energy.

- Determine the cost of the preventable or corrective action.

- Divide the cost of electricity lost by the cost of the corrective action. If you get a positive number, then it is a go.

Two methods will be shown:
Method 1. The traditional
Method 2. For each item, use energy waste, and using more energy as criterion in a criteria evaluation matrix, evaluate them as low (1 point), medium (2 points) and high (3 points). It has been determine that two mediums, one medium and one high or two highs are all cost effective.

Method—Traditional
1.   Estimated savings potential = hours times cost of energy
2.   Estimated countermeasure (pm activity) cost = labor hours times cost of labor per hour + materials estimate
3.   Savings/cost = if a positive number, then it is cost effective

<u>Traditional Example:</u>
Given: Preventative maintenance time required: 1.5 hours every six months. Material cost is $40 each time.

Craft is mechanical that is charged at $32.50 an hour.

kWh costs 8 cents per kWh. Wasted energy potential is 8,000 kWh in a year.

Calculation: Labor cost = 3 hours per year × $32.50 per hour = $97.50 per year. Material cost per year is $80. Electricity potential savings = 8000 kWh × .08 = $640 a year. Cost effective index = $640/$97.5 + $80 = 3.6

If a positive number above one occurs as an answer, it is cost effective to do the maintenance.

Method 2—Criteria Matrix-Method 2

Equipment/Component	Energy Waste Mode	Using More Energy	Countermeasure Need Indicator	Is Cost Effective?
Economizer	Seized or Broken M (3)	Gets stuck in the full open position & can use extensively more energy. H (5)	3x5=15	Yes. Assessed a M and a H.

Figure 6-10. Criteria Matrix

The ECM flow chart process does not include this latter step since often the electricians do this after the other decisions and actions have been done. However, it can be easily added if desired.

To figure out how much your devices cost to run, see these formulae:
•   Watts = Amps * Volts

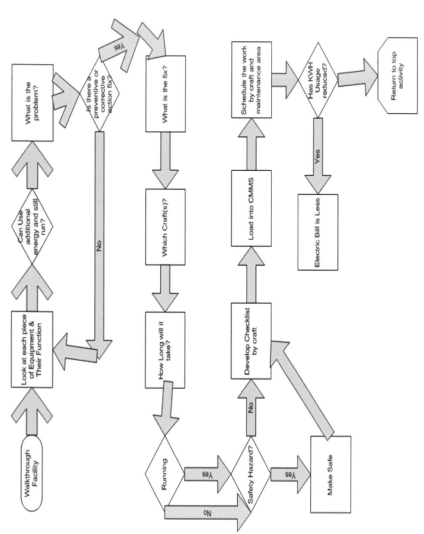

Figure 6-11. ECM Flow Chart

Kilowatts = Watts / 1000
Kilowatt-hours = kilowatts * hours used
Cost = kilowatt-hours * cost per kilowatt-hour
Or
* Cost = (((Amps * Volts) / 1000) * hours used) * cost per Kilowatt

Frequency, Craft and Time

Next, determine the frequency, craft and time.

Phase 1	Idendity the potential item
Phase 2	Test to determine any problem
Phase 3	If selected, determine what maintenance, repair or replacement is necessary
Phase 4	Ensure ECM is cost effective
Phase 5	Determine craft, frequency and labor required

Figure 6-12. Frequency, Craft and Time

Determine the preventative or corrective task
* Determine the frequency the task should be accomplished.
* Identify the craft to accomplish the task.
* Estimate to the best possible accuracy the time to complete the task with one or two people performing. (State additional people if needed—otherwise, one person for the time period is estimated.)

Maintenance Checklist and Computerized
Maintenance Management System (CMMS)

Maintenance checklists are developed and then put into CMMS to schedule the work and to have a maintenance history.

The maintenance hierarchy starts with the maintenance policy. It is a high-level document approved by management that spells out the maintenance goals, expected benefits, main components, and other policy-related items. Next is the maintenance strategy. It spells out actions or activities that will be followed to achieve the maintenance policy. The third component is the maintenance program. It includes the total maintenance planned for the facility and equipment. Next, is

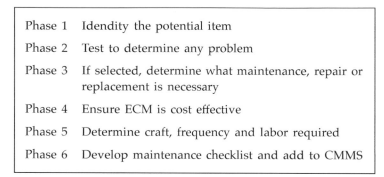

Phase 1	Idendity the potential item
Phase 2	Test to determine any problem
Phase 3	If selected, determine what maintenance, repair or replacement is necessary
Phase 4	Ensure ECM is cost effective
Phase 5	Determine craft, frequency and labor required
Phase 6	Develop maintenance checklist and add to CMMS

Figure 6-13. Maintenance Checklists

the maintenance checklist that includes a list of all maintenance tasks, frequency, and by what craft. The checklist enables a short-term or interval schedule to be developed that includes the work needed to be accomplished during a week or one shift.

Maintenance Policy

Guidance: Explain what the corporate commitment to maintenance is and management's expectations.

Example: ABC Inc. is committed to maintaining its equipment and facilities to ensure their functionality and ABC Inc.'s mission accomplishment. Operator's maintenance will be accomplished when equipment is started and shut down. Preventative and predictive maintenance will be scheduled and accomplished on time. Energy centered maintenance will be incorporated within the preventative maintenance program. Corrective maintenance will be accomplished when needed. Facility maintenance such as painting, roof repair, and lighting changes will be accomplished periodically when needed.

ABC Inc. strongly believes the maintenance programs will 1) Reduce equipment failures, 2) Reduce energy consumption and costs, 3) Reduce down time, 4) Improve productivity and 5) Increase equipment and facilities life.

Maintenance Strategy

Definition: A plan of action or policy designed to achieve a major or overall aim. A high-level plan to achieve one or more goals under conditions of uncertainty. A strategy determines how an end (goal) will

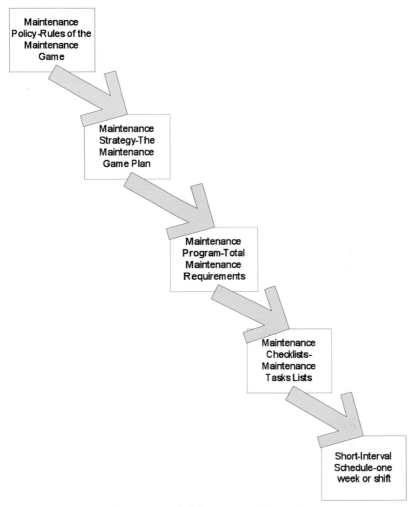

Figure 6-14. Maintenance Hierarchy

be achieved by the means (resources).

Strategy: Add ECM to the maintenance program by incorporating it with the preventative and predictive maintenance program.

Maintenance Program:

Guidance: Includes total maintenance program.

Total maintenance program = operator maintenance + PM + predictive maintenance + ECM + corrective maintenance + facility maintenance

A maintenance checklist summarizes each approved preventive or corrective maintenance activity onto one sheet and provides essential information for the activity to use in scheduling the work. Figure 6-1 shows the data entered on a checklist. Often a facility will develop a map of their facility divided into areas of work to be done. Normally, each area has about the same amount of work to be scheduled except where one craft may have a large concentration such as instrument and control (I&C). The areas are numbered alphabetically starting with A.

On the maintenance checklist the first item is the preventative or corrective maintenance task such as inspect and repair all switches pertaining to the boiler and HVAC. (The tasks could be broken down to individual switches and using the switch number.) Next is the frequency which is once a year. The area is "C." The time is 30 minutes per switch. The craft is electrical. If sensors are made apart of this activity, probably I&C would check and fix. Next is the safety item. Can the item be running when checked or should the power be turned off? In this case the inspection is with power on and the repair or replace is with the power off.

The preventative or corrective maintenance items are best done for one of the three crafts (mechanical, electrical and I&C) and then summarized by areas—all A's together, all Bs, etc.

• Integrate into CMMS (computerized maintenance management system).

• Create checklists for each craft that shows what has to be done by whom, when, time needed, whether a safety hazard is involved or not, and whether the equipment needs to be running or not.

• A rule of thumb: "For each maintenance period for a craft, do not let the time to accomplish a checklist be more than 80% of the total hours in a maintenance period."

Format is:

PM	Task	Freq	Area	Time	Craft	Safety	Run

CMMS is sometimes called CMMIS with the I being "Information" which describes what it is. It is a computerized maintenance management system that contains an organization's maintenance activities. The IS stands for information system. It is also referred to as the organization's facility maintenance program. A CMMS software package maintains a computer database of information about an organization's maintenance program and operations. This information is intended to help improve maintenance workers' efficiency and productivity (for example, determining which equipment requires maintenance—what and when and which supply rooms contain the parts or materials needed for a specific work order) and to help management make better decisions (for example, comparing the cost of equipment failure versus preventive maintenance for each machine). CMMS data may also be used to show regulatory compliance and compliance with an ISO requirement.

The following identifies what the CMMS should be able to do:

• Address all resources involved in maintenance.

• Maintain maintenance inventory and store room location.

• Record and maintain work history of all types of maintenance PM, predictive, emergency, and corrective.

• Include work tasks and frequencies for each craft.

• Effectively interface and communicate with related and supporting systems ranging from work generation through work performance, evaluation and performance reporting.

• Provide feedback information for analysis and decision making.

• Reduce costs through effective maintenance planning and execution.

A modern CMMS meets all these requirements and assists the facilities maintenance manager with, planning, scheduling, control, performance, evaluation, and reporting. Such a system will also maintain historical information for management use and provide meaningful maintenance metrics. Therefore, CMMS provides a work order system, asset management, inventory/purchasing, PM management, work scheduling, and management reports. CMMS is common in the manufacturing industry, facilities both government and civilian, fleet, service providers, oil and gas and other industries. CMMS software packages can be either web based (they are hosted by the company selling the product on an outside server), or LAN based, (The organization buying the software hosts the product on their own server.)

CMMS improves mechanic's or a technician's wrench time, enhances spare parts inventory and streamlines procurement of parts and materials. CMMS benefits are:

- Easy to use

- Quick to implement maintenance program

- Minimizes downtime and improves productivity

- Provides maintenance records and history that serves as compliance proof.

CMMS provides preventative maintenance scheduling, work orders, work or service requests, inventory control, predictive maintenance, maintenance reports, and other functions. The CMMS user interface allows for a really quick setup, easy data conversion, and the vendor provides training for the users.

CMMS vendors claim CMMS reduces costs and asset downtime, and increases productivity in less than a month. They claim it will extend the life cycle for your facility, decrease your liability, and lower your operating costs without any large upfront investment.

Provides online planned maintenance scheduling that helps generate, schedule, and manage recurring tasks which is the heart of ECM. Some systems allow sending work order information to maintenance crews in the field, enabling them to receive and complete tasks away from their shop. CMMS is very good at scheduling jobs, assigning personnel, ear marking materials, recording maintenance costs, and tracking information such as the cause of the problem (if any), downtime that occurred (if any), and suggestions for future action. The CMMS schedules preventive maintenance based on maintenance plans. Different CMMS software packages use different techniques for highlighting when a pm job or task should be performed.

CMMS keeps track of preventive maintenance jobs, including step-by-step instructions. This action is critical for ECM to be successful. The maintenance information process or loop is shown in Figure 6-15.

CMMS provides an online work order management system that streamlines your work order process, including work request generation, progress and completion status tracking and reports, and reporting of essential maintenance management data and information.

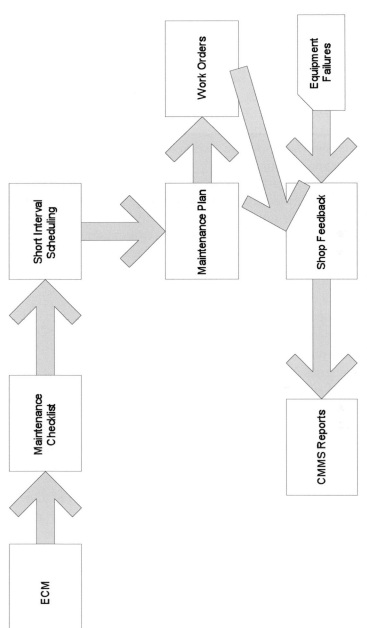

Figure 6-15. Maintenance Information Process

The maintenance checklist and short-term or interval scheduling were discussed above. The schedule is part of the facilities and equipment maintenance plan. Work orders are generated to cover what should be done, where, craft, whether equipment can be running or not and if any safety precautions are necessary. Work order information flows into a shop or craft summary that feeds information to CMMS reports. The CMMS reports enable management to determine PM and ECM effectiveness, make adjustments to the preventative measure or frequency, and adjust scheduling and workload. The maintenance information system provides an excellent audit trail including compliance to schedule information.

Reports can be generated by what can be sorted. With work order numbers and equipment numbers and others, there are many possibilities. The work order numbers can be specified to distinguish between preventative and predictive maintenance from energy centered maintenance. Some reports that are meaningful are:

1. Labor hours by area and shop (craft).
2. Material costs by area and shop (craft).
3. Equipment failures by month and area.
4. Equipment downtime calculated by week or month.
5. Emergency response time.
6. Percent of schedule completed on time.
7. Percent availability of materials.
8. Percent of scheduled facility maintenance accomplished.
9. Amount of backlog by category by month.
10. Amount of emergency maintenance and/or corrective maintenance man-hours, materials, and total cost.

There are many other possibilities. Customize them to your needs. There are many CMMS vendors with software packages that can provide the functions and benefits described above. If you are interested in purchasing a CMMS system, go online, put CMMS in your search bar, and you will get the vendors. Some have videos of their system. Develop a specification for what you would like to have and compare with the different possibilities until you find a close match at a price you can afford. Many large facilities including military and federal agencies already have a CMMS or a similar maintenance program or system available for their use.

Schedule the Work

The last phase is to schedule the work.

Phase 1	Idendity the potential item
Phase 2	Test to determine any problem
Phase 3	If selected, determine what maintenance, repair or replacement is necessary
Phase 4	Ensure ECM is cost effective
Phase 5	Determine craft, frequency and labor required
Phase 6	Develop maintenance checklist and add to CMMS
Phase 7	Schedule work and monitor

Figure 6-16. Schedule the Work

Short interval scheduling is normally used to schedule the work when it is designated to be done, by craft, and man hours allocated in the craft's work time. Most CMMS incorporate a form of short interval scheduling. Figure 6-17 Schedule provides a typical schedule for ECM.

ECM Schedule

Operation Unit	Craft	Wk 1	Wk 2... Wkn
Area 1	Mech		
	Elect		
	Inst		
Area 2	Mech		
	ECM		
	Inst		

Figure 6-17. Schedule

First, we identify ECM, determine whether it is cost effective or not, enter it into a CMMS, and then schedule it to be inspected or repaired. These activities have been outlined. Once a full ECM is completed, the energy metrics such as kWh consumption or kWh intensity should reflect a reduction and the organization moving closer to meeting its energy reduction goals.

In other words, do these three actions:

☐   Put preventative maintenance system (including energy centered maintenance), into the CCMS.

☐   Schedule the work on work orders

☐   Post when the item has been checked and "fixed."

The criteria to determine if an item is to be included in ECM are:

•   The equipment or component does not fail. It still meets its function.

•   Something causes the equipment or component to start using more energy than is normally required.

•   The cost to prevent the equipment or component to use excessive energy is less than the additional kWh cost.

USING RELIABILITY THEORY

Gather reliability information for analysis and decisions—start with fact sheet & ECM analysis.

Reliability information is helpful in analyzing and developing the proper course of action for each ECM item.

Reliability is the probability of performing without failure, a specific function under given conditions for a specified period of time.

The elements are:

☐   Probability

☐   Failure

☐   Conditions

☐   Functions

☐   Time

The data needed are:

☐   Determine equipment or components that may continue to function while using more than normal or expected energy.

☐   Determine the equipment's or component's function.

☐   Describe what has to happen for it to still function while using excessive energy.

☐   Describe the energy waste mode and assess its probability of occurrence by high (5 points), medium (3 points) or low (1 point).

❏   Describe how severe the energy waste (kWh for electricity, cubic feet for gas) for a 24-hour period when it occurs (high, 5 points), (medium, 3 points), (low, 1 point).

❏   Multiply the probability of occurrence by the severity to get a countermeasure need index (the higher, the more need).

❏   Determine the corrective maintenance or preventative maintenance needed to lower or prevent the energy waste mode from occurring.

Using the failure modes effect analysis (FMEA), an energy effect modes analysis (EEMA) can be developed and the concept used in ECM.

Table 6-1. Excessive Energy Modes Effects Analysis (EEMEA)

Equipment or Component (A)	Function (B)	Excessive Situation (C)	Excessive Energy Mode (D)	Probability of Occurrence (1 to 5 with 5 being most probably will occur) (E)	Severity (1 to 5 with 5 being very severe in energy loss) (F)	Counter-measure Need Index (G) (E) X (F) =(G)
1. Fans	To provide air flow for cooling	Fans slowed due to needing lubricated	Needs Maintenance	1	3	3
2. Light Bulbs	To provide lights	Light presented is less due to dirt film	Needs Cleaning	3	2	6

The EEMEA can be used to determine the probability and severity of each possible item and its priority in the need to accomplish. Also, three additional columns can be added at the righthand side to add the corrective or preventative action that needs to be done to prevent the energy waste, list the frequency and labor hours used, and name the craft that needs to do the work. Later, a simplified template will be used in lieu of the EEMEA. The reader can select either for their use if deemed helpful.

The EEMEA could be useful in establishing priorities for major projects since the cost is much higher than for ECM preventative maintenance activities where the cost is very inexpensive, normally consisting of 1-2 hours with low material costs.

APPLYING THE ECM PROCESS—GETTING STARTED—
THE STEPS TO TAKE

Getting Started

In applying the first four phases, it is helpful to use the steps outlined in each phase. Walk through the facility and identify the equipment. Then decide if components of the equipment are better to track and monitor. The equipment, component or items identified in most walkthroughs are listed below:

1. Fans, bearings, and belts

2. Leaks—doors, windows, plumbing, outlets, outside air intakes, HVAC, cabinets, and ducts on rooftops

3. Economizers

4. HVAC and condenser coils

5. Boilers

6. Motors

7. Lights

8. Air compressors

9. Data center equipment

10. Computers and monitors

11. Imaging equipment

12. Energy conservation measures

The first item identified was fans, bearings and belts. Now, step 2 should be accomplished.

Step 2. Develop a short fact sheet that explains what the function is, how the item can use excessive or waste energy and the cause that can make this happen. The short fact sheet developed is shown as Figure 6-18 Fact Sheet.

Figure 6-18. Fans, Bearings, and Belts Fact Sheet

Inspect fan blades, bearings, and belts at least annually to prevent failure. The fan blades should be cleaned, bearings should be examined for adequate lubrication, and belts should be adjusted and if needed, changed.

Next, identify their function and other ECM data as shown in the cover sheet for fans, bearings, and belts below.

Fans, Bearings, and Belts

Inspect fan blades, bearings, and belts at least annually to prevent failure. The fan blades should be cleaned, bearings should be examined for adequate lubrication, and belts should be adjusted and if needed, changed. The fans' function is to cool area for added comfort. The ECM analysis for fans is shown in Figure 6-19.

A simple sheet to reflect cost effectiveness is shown as Figure 6-20.

Identify Equipment or Component	ECM Identifications	Root Cause
Fans	**Function**: To cool area for added comfort. **Energy Waste**: Fan not working and AC has to increase load. Fan moving at a slower speed due to control switch not working properly. **Energy Waste Cause**: Worn	Improper maintenance due to inadequate lubrication Fan has worn out in the switch area.
	Prevention/Corrective Action: Inspect the fan to see if anything is not working properly. Speed? Stability? Replace switch if needed and lubricate fan. The potential KWH Lost is higher than the preventative action	
	Task Frequency/Craft: Semi-Annual / Electrical/C.E. Yes	

Figure 6-19. ECM Analysis—Fans

Equipment/ Component	Energy Waste Mode	Using more Energy	Countermeasure Need Indicator	Is Cost Effective?
Fan	Worn or switch not working properly Medium (3)	AC has to work more since fan is at a slower speed Medium (3)	(9)	Yes. Since two Ms were assessed.

Figure 6-20. Cost Effectiveness Determination—Fans

Economizers

A lot of air-conditioning systems use a damper vent called an economizer to draw in cool outside air when it is available to reduce the need for mechanically cooled air. The linkage on the damper, if not regularly checked and lubricated, can easily seize up or break. An economizer that's stuck in the fully open position can increase a building's annual energy bill as much as 50% by allowing hot air in during the air-conditioning season and cold air in during the heating season.

HVAC

Change filters when needed—do on a schedule. Change air filters every one to three months. Air conditioners that are located next to highways or construction sites, or that are using an economizer will also need more frequent filter changes. Clean the condenser coils.

Condenser Coils

Clean the condenser coil annually if it has any debris. Check quarterly or semi-annually to see if cleaning is needed. A dirty coil that raises condensing temperatures by 10° Fahrenheit can increase power consumption by 10% for the unit. Condenser coils should be checked for debris on a quarterly basis and cleaned.

Lights

Look for the blue and white Energy Star® label on compact fluorescent light bulbs (CFL) or light-emitting diode (LED) bulbs. They

Identify Equipment or Components	ECM Identifications	Root Cause
Economizer	**Function**: Draw in cool air from outside. **Energy Waste**: Gets stuck in the full open position & can use extensively more energy. **Energy Waste Cause or Mode:** Seized or Broke	Linkage on damper seized or broke.
	Prevention/Corrective Action: Check and Lubricate the linkage on the damper. The preventative action cost is much less than the potential KWH lost. **Task/Frequency/Craft**: Inspect and lubricate linkage/Quarterly/Mechanical **Time**: 15 minutes for each C.E. Yes	

Figure 6-21. ECM Analysis—Economizers

Equipment/Component	Energy Waste Mode	Using More Energy	Countermeasure Need Indicator	Is Cost Effective?
Economizer	Seized or Broken M (3)	Gets stuck in the full open position & can use extensively more energy. H (5)	3x5=15	Yes. Assessed a M and a H.

Figure 6-22. Cost Effectiveness Determination-Economizer

use up to 75% less energy than standard incandescent bulbs. Place on your preventative maintenance program to clean dirt & grease from light bulbs and fixtures to increase lighting output levels by 10%.

Identify Equipment or Component	ECM Characteristics	Root Cause(s)
HVAC	Function: Provide cool air or hot air and circulate it through the facility.	
	Energy Waste: A dirty filter makes the system to work harder. When the system works harder it uses more electricity or energy.	
	Energy Waste Cause or Mode: Clogs	
	Prevention/Corrective Action: Change filter per mfg. recommendation or sooner. Cost is much less for the corrective action than the KWH Wasted cost. Also, a health issue if not done.	Dirt
	Task/Frequency/Craft:	
	Filter-Mech	
	Time: 10 minutes each filter /C.E. Yes	

Figure 6-23. ECM Analysis-HVAC

Motors

Look for motors running excessively. Lubricate them. Good places to start are those associated with boilers and chillers.

Boilers

Include an item for treating makeup water to prevent equipment damage and efficiency losses. If not, build-up inside the tank will result and decrease heat transfer to the water and necessitate more frequent blowdown. This wastes both water and energy. Check the air fuel ratio

Equipment/Component	Energy Waste Mode	Uses More Energy	Countermeasure Need Indicator	Is Cost Effective?
HVAC	Filter clogs High (5)	A dirty filter makes the system to work harder. When the system works harder it uses more electricity or energy. High (5)	HxH=5X5=25	Yes. Two H's were assessed.

Figure 6-24. Cost Effectiveness Determination-HVAC

to ensure that the combustion process is operating efficiently. This will ensure excellent combustion efficiency.

Air compressors

Check hoses and valves for leaks regularly, and make any repairs if needed. A poorly maintained system can waste between 25 and 35 percent of its air (due to leaks alone). Cleaning intake vents, air filters, and heat exchangers on a schedule will increase both equipment life and efficiency.

Leaks—Building Envelope and Seals

Air infiltration, air leaking out through gaps around doors on the receiving and loading docks or any doors entering the facility are an energy waste. At least annually, check and repair gaps in door seals, and ensure all employees keep the doors closed. Inspect all windows to

Identify Equipment or Component	ECM Characteristics	Root Cause(s)
Condenser Coils	Function: Condenses the air	
	Energy Waste: Can raise temperature by 10 degrees F which increases energy use by 10% for the unit	
	Energy Waste Cause or Mode: Over heats	
Condenser Coils	Prevention/Corrective Action: Clean the coils. Cost of Cleaning is much less than the potential energy loss. Task/Frequency/Craft: Clean the condenser coils annually-Mech. Time: Varies by size of Unit C.E. Yes	

Figure 6-25. ECM Analysis-Condenser Coils

see if leaks are occurring. Check to see if any insulation can be added in the roof or outside walls to reduce energy use.

Leaks—HVAC Equipment

On a quarterly basis, cabinet panels and ducts on rooftop HVAC equipment should be checked for leaks. A structural inspection should be made to ensure that the units are secure, with all screws in place. Annually, inspect all access panels and gaskets—particularly on the supply-air side, where pressure is always higher. Just one leak in an HVAC rooftop unit can cost $100 per unit per year in wasted energy.

Identify Equipment or Component	ECM Characteristics	Root Cause(s)
Office Lights	Function: Provides light to office workers.	
	Energy Waste: Decrease lights outputs level while using same energy.	
	Energy Waste Cause or Mode: Covered in dirt and grease	
	Prevention/Corrective Action: Clean light and fixture. Cost for cleaning is insignificant with light gained.	
	Task/Frequency/Craft: Clean Annually/Mech	
	Time: 1 .5 minutes per light fixture/C.E. Yes	

Figure 6-26. ECM Analysis-Office Lights

Outside-air Intake Controls

A lot of facilities have rooftop units for heating, ventilation, and sometimes cooling. Many are equipped with exhaust fans that bring in outside air for ventilation. Set these to run only when spaces are occupied. (It is not uncommon to see them run 24/7.)

Energy Centered Measures (ECM)

The next two are energy centered measures (ECM). They are computers and monitors and imaging equipment. Both are great op-

Identify Equipment or Component	ECM Characteristics	Root Cause(s)
Sensors/switches for Motors	Function: To provide power to circulating pumps.	Switch failed due to excessive volts in system.
	Excessive Energy Waste: Motor runs constantly instead at variable times when needed.	
	Excessive Energy Cause or Mode: Switch to turn motors on and off has failed.	
Sensors/switches for Motors	Corrective/Preventive Actions: Inspect all switches/sensors/and controls for the boiler and HVAC systems.	
	Craft/Frequency/Time: I &C, once a year, and Time is 8 Hours.	
	Cost Effective: Yes.	

Figure 6-27. ECM Analysis-Sensors/Switches for Motors

portunities to reduce energy consumption and cost of electricity. How to implement these measures will be discussed further in Chapters 10 and 11 respectively.

CPUs and Monitors

Figure 6-33 gives the computers and monitors fact sheet then followed by the ECM analysis.

Imaging Equipment

Figure 6-35 gives the imaging equipment fact sheet followed by Figure 6-36 ECM Analysis.

Identify Equipment or Component	ECM Characteristics	Root Cause(s)
Boiler	Function: To generate steam or hot water.	
	Energy Waste : Build-up from Make Up Water.	
	Energy Waste Cause or Mode: Buildup inside of the Boiler	
	Corrective or Preventative Action: Check and treat makeup water. Check air fuel ratio	
Boiler	Craft/Frequency: Mech Daily or Weekly based on past experience	
	Time: 30 minutes	
	Cost Effective: Yes	

Figure 6-28. ECM Analysis-Boilers

Energy Waste Types
1. Running when it should be turned off (motors, computers, and monitors).
2. Having to work harder (HVAC).
3. Worn (fans, belts).
4. Leaks (doors, windows, air compressors).
5. Stuck in the open or closed position (switches, motors, economizers).
6. Underperforming (dirt or grease on lights).
7. Over performing (imaging machines).
8. Overheating (condensers).
9. Waste build-up (boilers, filters).

Identify Equipment or Component	ECM Characteristics	Root Cause(s)
Air Compressors	Function: Provide air upon demand.	
	Energy Waste: Hoses and valves leaks.	
	Energy Waste cause or mode: Excessive leaks	
	Corrective/Preventative Actions: Check hoses and valves and replace if leaking.	
Air Compressors	Craft/Frequency: Mech Monthly	Worn with Use
	Time: 30 minutes per compressor	
	Cost Effective: Yes	

Figure 6-29. ECM Analysis-Air Compressors

10. Incorrect settings (exhaust fans).
11. As organizations implement ECM, this list could get longer. In this chapter, all of the energy waste modes have been shown in an example.

Energy waste is everywhere. ECM will help eliminate or minimize it and do it cost effectively.

Identify Equipment or Component	ECM Characteristics	Root Cause(s)
Windows & Doors	Function: To allow access to facility and to have some	Weather and aging
	day light enter the facility.	
	Energy Waste: Leaks in window seals and around door let cool air out and hot air in or vice versa.	
	Energy Cause Mode: Leaks/Not sealed	
Windows and Doors	Corrective/Preventative Action: Inspect for Leaks and seal when detected.	
	Craft/Frequency: Mech & annually	
	Time: one hour for a door and 45 minutes per window if both detection and repair are accomplished Cost Effective: Yes	

Figure 6-30. ECM Analysis-Windows and Doors

Identify Equipment or Component	ECM Characteristics	Root Cause(s)
HVAC Cabinet s and ducts-rooftop, access panels and gaskets	Functions: To house air conditioning or heat units and to distribute air to facility.	Weather and age
	Energy Waste: Leaks	
	Energy Waste mode: Holes allowing leaks.	
	Corrective/Preventive Cause: Inspect for leaks. Make sure they are tied down and secure.	
	Craft/Frequency/Time: Mech/annually/1 hour inspection. To fix will take different times depending on what and severity of damage.	
	Cost Effective: Yes	

Figure 6-31. ECM Analysis—HVAC Ducts

Identify Equipment or Component	ECM Characteristics	Root Cause(s)
Exhaust Fans	Function: Bring outside air into facility	Sensor, setting, or switch malfunction
	Energy Waste: Should be set to run only when facility is occupied.	
	Energy Waste Mode: Running when it should be off.	
Exhaust Fans	Corrective /Preventive Action: Inspect and fix ones running all the time.	
	Craft/Frequency/Time: Mech/semiannually/2 hours average	
	Cost Effective: yes	

Figure 6-32. ECM Analyst—Exhaust Fans

Computers(CPUs) & Monitors

Computers and Monitors are often left on when not in use. This occurs off shift, weekends, holidays , and even vacations. The Monitors and CPUs can be set to turn off after a certain time(15-30 minutes) and save energy and money(around $75 annually for each). This action is called IT Power Management Program. (See Chapter 14)

Figure 6-33. CPUs & Monitors Fact Sheet

Identify Equipment and/or Component	ECM Characteristics	Root Cause
CPUs and Monitors	Function: To accomplish administration or mathematical functions, to use internet and develop and receive emails, shop and perform numerous other activities.	
	Energy Waste: Uses electricity when they are idle and no one is using them.	Have not trained employees on Power Management nor implemented an IT Power Management Program.
	Energy Mode: Running when it should be turned off or hibernate or in sleep mode.	
CPUs & Monitors	Corrective/Preventative Action: Implement IT Power Management	
	Craft/Time: Data Processing Personnel Time will varying on number of computers & Monitors	
	Cost Effective: Will save about $75 for each yearly.	

Figure 6-34. ECM Analysis Computers and Monitors

Imaging Equipment(Copiers, Printers, and Others)

Imaging machines can copy numerous pages of items that may not be necessary if duplex copying or other "Reduce Office Paper" countermeasures are employed. Organizations have saved as much as 30-50% reduction in office paper use in 1-2 years. Less use of imaging equipment saves energy.

Figure 6-35. Imaging Equipment Fact Sheet

Identify Equipment and/or Components	ECM Components	Root Causes
Copiers, Printers, Multifunction Machines	Function: Copying paper	
	Energy Waste: Energy used to copy papers that are un-necessary.	Not using Duplexing and no reduce office paper program implemented
	Energy Mode: Over Performing or Over Production	
Copiers, Printers, and Multifunctional Machines	Corrective or Preventable Action: Implement a Paper Reduction Program	
	Craft and Time: Energy Team. Time is minimal.	
	Cost Effective: Yes. Can reduce paper cost by 50% in 1-2 years and save tons of paper and reduce paper purchasing cost.	

Figure 6-36. ECM Analysis-Imaging Equipment

Chapter 7

Energy Reduction Deployment And ISO 50001
Energy Management System

INTRODUCTION

Energy centered management system supports ISO 50001 Energy Management System (EnMS). Both have as an objective/goal to manage energy. ECMS focuses on reducing energy consumption. ISO 50001 EnMS focuses on on continuous improvement in energy management.

ENERGY REDUCTION DEPLOYMENT OUTLINE AND
WHAT IS NEEDED TO CONFORM WITH ISO 50001 ENMS

Five management components comprise the ECMS method or system—all leading to energy waste reduction and energy consumption and cost reductions. They are: ECP&D, ECW, ECO, ECP, ECM

ECP&D
— Top management commitment
— Appoint energy champion
— Set corporate goal/objective
— Approve energy policy and communicate
— Energy champion establishes cross-functional energy team
— Do baseline study and establish energy performance indicators

ECW
Find Energy Waste
— Conduct management/employee energy brainstorming sessions

— Establish energy functional teams
— Establish a walkthrough team
— Team composition must be energy knowledgeable
— Identify significant energy users
— Identify energy waste that can be fixed with corrective or pre-ventative maintenance
— Identify potential projects that will reduce energy use and consumption
— Walkthrough should cover all major areas of possible energy waste
— Provide energy awareness training
— Develop an energy conservation program
— Develop an energy plan and communicate
— Must be compelling and complete
— Set the actions for next 1-3 or 1-5 years

ECO

— Energy team establish objectives and targets (O&T)—feasibilities study—improve/maintain objectives
— Assign responsible person for each O&T
— Responsible person with help of energy team develops the action plan
— Responsible person implements and uses project management to ensure success of objective and target

ECP

— Develop project specifications with projects with good payback and funded by top management
— Advertise and award contract if a good bid is received
— Manage the contract until acceptably completed
— Develop and management re-commission effort if approved by energy team, energy champion and top management
— Secure an energy services performance contract or utilities energy service contract.

ECM

— What it is
— Objective
— ECM vs. RCM

— Savings/cost avoidance
— ECM's aim
— ECM components
— ECM process and four phases
— Cost effectiveness
— FMECA turned into EEMEA
— General findings on cost effectiveness
— Applying the process
— Summary sheet
— ECM template and analysis
— Maintenance check list
— Computerized maintenance management system (CMMS)

ISO 50001 ENMS WOULD REQUIRE ADDITIONS

Energy reduction deployment would need the following additions if the decision is to turn it into an EnMS (ISO 50001):

- Legal and other requirement and annual evaluation to ensure compliance. List all the federal, state, city, county and local energy and energy related legal requirements and any other requirements of a higher headquarters or the facility itself.

- Developing operational controls and implementing them. Determine what controls are in place or can be developed to mitigate the significant energy users getting worse by using additional energy.

- Documenting requirements and control of records. Have a central system, either an organization's operational system or a Share Point system, to document what happens in planning, developing, implementing and maintaining the EnMS. Files should include as a minimum:

1. The corporate goal and energy policy.

2. Appointment of the energy champion and his roles and responsibilities.

3. The energy team charter, energy team members and roles and responsibilities such as team leader, facilitator, and note taker.

4. Energy team meetings agendas, meeting minutes, and sign-in

sheets by month and year.

5. Objectives and targets and action plans by year, open and completed.

6. Training including awareness and energy conservation.

7. Communications plan and examples of internal and external plans

8. Measurements and a monitoring list of items that need to be tracked and monitored by the energy team.

9. Operational controls

10. Emergency or contingency plans.

11. Self inspections and internal audits.

12. Nonconformities, deficiencies, and corrective actions.

13. Management reviews agendas, minutes, and presentations given.

- **Contingency planning** to include backup energy supply.

- **Monitoring-measurement** is a part of ECMS—the performance indicators and anything that needs to be tracked such as energy bills, significant energy users, office paper reduction, etc. This is recommended in ERD but not in detail as required by the standard.

- **Implementing nonconformities, corrective and preventative actions**. Include any corrective actions reports, self-inspection or internal-audit documentation and findings and recommendations. Recommended in ERD but not mandatory.

- **Management reviews** with designated inputs and outputs. Include the agendas, minutes, sign-in sheets, presentations, action items, and examples of continuous improvement.

  Inputs to meetings should include 1. Environmental policy; 2. Performance measures (progress toward goal); 3. Objectives and targets and action plans status; 4. Self inspections or internal audits; 5. CARs (corrective action reports); 6. Energy plan status and future actions; 7. Any issues or barriers to progress; 8. Any projects needing funding such as metering plan and lights replacements.

  Outputs of the meetings should include: 1. Actions items that need to be done and time allowed for completion; 2. Recommended changes to energy plan, environmental policy, and objectives and targets; 3. Recommendations for improvement; 4. Approvals of objectives, targets and future plans; 5. Approval of projects or recommendations for changes to make more cost effective.

All of the above are excellent actions and probably should be done anyway. Management review is a part of the energy centered management system as well as measurement. Here they are expanded some to meet the standard. The ISO 50001 Energy Management System (EnMS) has no end; it is a journey and there will always be a need for the energy team to meet. The frequency of the meetings may change from monthly to quarterly, but the need to meet will always exist. The energy centered management system can exist to meet a corporate goal and then be dissolved, unless a new more compelling goal is established by management; or it can be extended by management as a continuous improvement system. It would be a good idea to keep the energy team but their meetings and involvement would be like a housekeeping effort. The energy centered management system, if desired by management, can be made a continuous journey and go through the stages—planning, development, implementation, maintaining/sustaining like an EnMS does. The first stage after achieving the corporate goal would be the maintenance stage. Here the energy team would meet less, have fewer objectives and targets, and do only the absolutely mandatory actions.

SUPERIOR ENERGY PERFORMANCE

SEP is a certification and recognition program for facilities demonstrating energy management excellence and sustained energy savings. DOE, ANSI and ANAB industrial facilities and water/waste water facilities can pursue Superior Energy Performance™ (SEP) to develop a verifiable energy management system, save energy costs through operational and capital improvements, increase productivity, and improve competitiveness. In addition, facilities can earn the globally recognized, U.S. Department of Energy supported SEP certification to gain recognition for energy excellence.

SEP enables facilities to achieve continual improvements in energy efficiency while boosting competitiveness. In the future (2-3 years), the author believes commercial and government facilities may be included in SEP.

SEP Continuum

Figure 7-1 SEP Continuum shows that ISO 50001 EnMS must be implemented and certified before an organization can reach SEP for certification. The SEP continuum is explained in Figure 7-2 SEP Continuum Steps.

As an organization considers their facility's readiness for energy management, they can use the strategic energy management checklist to conduct a high-level assessment of their facility's energy management program and identify steps you can take to move to the next level. This simple checklist can help you assess whether your facility should consider ISO 50001 and Superior Energy Performance™ (SEP) as a next step, or if you would benefit more from foundational energy management programs, such as the Environmental Protection Agency's ENERGY STAR for Buildings and Plants.

To use the checklist, select the energy management activities that your facility has already implemented. The checklist then generates a report of ISO 50001 and SEP elements that have been completed and those that remain to be implemented. The result can also help you determine your position along the strategic energy management continuum and whether ENERGY STAR, ISO 50001 or SEP certification is an appropriate next step. The checklist is found online; just search for Superior Energy Performance. It covers the elements of ISO 50001 EnMS written in a question format so that whether it has been implemented can be confirmed. The areas covered are:

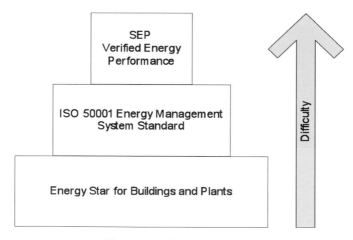

Figure 7-1 SEP Continuum

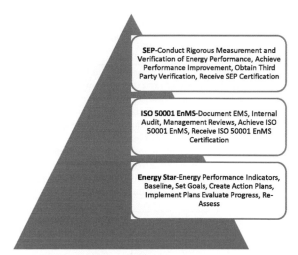

Figure 7-2 SEP Continuum Steps

- Energy performance indicators (EnPIs)
- Objectives and targets
- Energy action plans
- Competence training and awareness
- Communications
- Documentation
- Operational controls
- Design
- Procurement
- Monitoring and measurement
- Internal audits
- Corrective and preventive actions
- Management reviews
- Recognition

The additional SEP requirements to ISO 50001 are listed in Table 7-1.

Application Requirements Are:
SEP is open to industrial facilities and water/wastewater facilities. They could be private-sector and government-owned facilities. SEP is

Table 7-1. SEP Additional Requirements

ISO 50001 EnMS Element	SEP Additional Requirement
Management Responsibility	My organization has defined the scope of its EnMS as the facility, which includes the entire area occupied by the organization at a particular location. My organization's top management has ensured that the energy management team has taken into account each energy source consumed within the defined facility boundaries and the SEP facility-wide energy performance indicator (SEnPI) reflects each energy source[2] NOTE: SEP requirements were developed so that the organization manages each energy source within the facility and not just a subset of the energy sources. However, SEP does allow for a facility to exclude up to 5% of total energy consumption.
Energy Policy	None.
Energy Review- Analyzing Consumption Data	My organization collects energy data on each energy source that crosses the facility boundaries defined by the scope of the EnMS. My organization measures energy consumption at the facility's physical boundaries and accounts for at least 95% of the total energy consumption crossing the boundary of the EnMS[3] [3]For further information on boundary energy sources and consumption see the SEP M&V Protocol, sections 3.3.1, 3.3.2 and 3.3.3.
Objectives and Targets	None.
Legal and Other Requirements	My organization subscribes to the additional requirements of the SEP program which go beyond ISO 50001 and regularly evaluates compliance to those requirements as prescribed by ISO 50001, section 4.6.2. "Evaluation of compliance with legal requirements and other requirements". The SEP program constitutes an "other requirement" within ISO 50001
Energy Review- Significant Energy Users	My organization keeps up to date its list of facilities, equipment, systems and processes that account for the majority of energy consumption.
Energy Review- Metrics	My organization uses 12 months of data to develop the SEnPI to ensure that operating patterns over all seasons of the year are taken into account[5] My organization takes into account relevant variables that affect energy consumption and develops normalized models to establish the SEnPI and baseline. The baseline and SEnPI meet the statistical requirements as specified in the SEP M&V Protocol for Industry.
Energy Action Plans	For each action plan, my organization: Estimates expected energy savings, Implements action plans that achieve our SEP energy performance improvement threshold[7] Verifies the actual energy savings achieved, and Tracks the energy performance improvements in our action plans to check the energy performance improvement calculated by the SEnPI. This bottom-up sanity check confirms the energy performance improvement could reasonably have resulted from the action plans.
Competence and Awareness Training	For personnel working with SEUs, my organization keeps records of their training needs and when the actual training was delivered.

Communications	My organization communicates with the SEP Administrator as required and has established an external communication procedure for doing so. For external communication, see section 4.5.3 of ISO 50001 and section 4.5.3 of ANSI/MSE 50021-2013 which says, "The organization shall establish and implement a method for communicating with the SEP Administrator as part of its external communication process". For further information on the SEP Administrator, see SEP Certification Protocol, December 7, 2012, sections 3.4, 4.1, 4.2, 4.5 and 6.0. The SEP Certification Protocol can be accessed at: http://www.superiorenergyperformance.net/pdfs/SEP_Cert_Protocol.pdf
Documentation	My organization has a documented procedure for the control of records.
Operational control, Design and Procurement	None
Monitoring and Measurement	My organization monitors and measures its SEnPI to track its overall energy performance improvement.
Internal Audit	My organization's internal audits cover the SEnPI and energy performance improvement to ensure that we are meeting the requirements in the SEP M&V Protocol. My organization is pursuing the SEP Mature Energy pathway, and therefore, our internal audits cover the additional requirements in the SEP Best Practice Scorecard for Industry.
Corrective Action	My organization has a documented procedure for corrective action Note: ISO 50001 does not require a documented corrective action procedure but SEP does.
Management Review	Review of energy performance by top management within my organization also includes: Review of the SEP performance pathway (energy performance or mature energy), the SEP performance level (silver, gold, platinum) and SEnPI Making decisions concerning changes to the SEP performance pathway and SEP performance certification level.
Recognition	Recognition-My organization is certified to ISO 50001 and the SEP Program and has achieved the SEP recognition level - Silver, Gold or Platinum - for either the Energy Performance or Mature Energy pathway. An ANSI-ANAB accredited verification body (VB) has certified both the EnMS and the SEP program requirements.
	Recognition-My organization is certified to ISO 50001 and the SEP Program and has achieved the SEP recognition level - Silver, Gold or Platinum - for either the Energy Performance or Mature Energy pathway. An ANSI-ANAB accredited verification body (VB) has certified both the EnMS and the SEP program requirements.

only for facilities wanting third-party certification. There is no self-declared recognition by DOE, so they need internal audit as required by ISO 50001 and also the third-party audit by an SEP verification body.

They must meet ISO 50001 EnMS Energy Management Standard and meet additional performance requirements including attainment of established energy goals specified in ANSI/MSE 50021.

Applications are submitted to the SEP administrator. The company must contract a third-party ANSI-ANAB certified SEP verification body who provides certification of the company's body of work. Applicants may choose between the "Energy Performance Pathway" and the "Mature Energy Pathway" to reach one of these designations.

EnMS must conform to ISO 50001 EnMS standard and ANSI/MSE 50021. Then, energy performance improvement over 3 years after baseline must reduce energy consumption by silver 5%, gold 10%, and platinum by 15%.

The energy management system conforms to ISO 50001 EnMS and to ANSI/MSE 50021. Energy performance improved over 3 years after baseline. Energy performance improved at least 15%, 5 to 10 years after the baseline.

Chapter 8

Self Inspection and Internal Audits

INTRODUCTION

After an energy management system has been in place about a year, it is highly beneficial to conduct a self inspection or internal audit. Thereafter, self inspection should be done once a year around the same time. Internal audits every 3-4 years. Self inspections and/or internal audits, objectives and targets along with management reviews, are the major drivers for continuous improvement.

There are three types of self inspections. The first type is a "gap analysis." It is a formal way of identifying current situations and comparing them with desired conditions to determine if there are any "gaps." One or two people can perform the gap analysis. It can be accomplished even before an energy management system is implemented or anytime afterwards. Normally it is performed only one time and is not a periodically reoccurring tool for self inspections. Procedures, documents and records are reviewed, meetings are observed, and key people interviewed. A report showing the findings is written and discussed with management.

The second self-inspection method is accomplished either by one or two people involved in the organization's energy management system or by an energy team. The process is to identify the mandatory requirements that you want included in your EnMS. Next, identify the areas they cover. Put these factors into one or two words. These are called excellence factors. They are similar to key results areas (KRAs) and critical success factors (CSFs). These factors make it easy to identify deficiencies with areas or elements whose efficiency is mandatory, or whose requirements are deficient (either not done or done incorrectly). Reviews of documents along with interviews are the most common tools used in conducting the self inspection. This type of self inspection should be done annually and a report written, deficiencies corrected both using action plans or CARs (corrective action reports), and discussed at the next management review as one of the inputs.

The third method is the most often used by organizations that have a cross-function energy team. An audit checklist is developed that covers most of the key elements. This checklist is filled out by the energy team or an internal audit team (the same checklist used for both the self inspection and the internal audit). The checklist is completed element-by-element until the team's reviewers are satisfied enough to check concurrence (yes) or (no). This self inspection should be done once a year around the same time. The findings can be written on the checklist and no report required. However, if used to conduct an internal or external audit, a report is recommended. Self inspections are done yearly and internal audits every 3-4 years. The deficiencies or non-conformances are resolved through the CAR (corrective actions reports) process and verified that they were actually corrected.

GAP ANALYSIS

A gap assessment is recommended to assess whether your implementation plan (tasks, responsible persons, resources, timelines) is adequate and to identify those areas that are "gaps" to be filled to implement a satisfactory ISO 50001 system or energy centered management system (ECMS).

It will also enable your organization to determine whether there is an opportunity to integrate ISO 50001 or ECMS with any other management systems such as ISO 14001 Environmental Management Systems, ISO 9001 Quality Management Systems, OHMS 18000 Safety Management System, or build upon existing energy conservation efforts underway at your facility or facilities.

Gap analysis identifies and then compares the gap between an organization's actual performances against its potential or desired performance. In gap analysis, you typically list the organization's current situation, its desired future state, and then develop a comprehensive plan to fill the gap between these two states. This analysis can result in a lot of insights into an organization's current performance and show what has to be done to meet the desired future organizational performance.

Step 1. List all attributes you would like to see improved. For ISO 50001, the elements may be appropriate. For ECMS, the musts would be appropriate or the deliverables.

Step 2. Identify all the activities you desire to see improved or to possess in the future state. For ISO 50001 EnMS this would be the requirements and for ECMS the deliverables.

Step 3. Bridge the gap. Ask does there exist a gap between the current situation or state and desired state. Answer yes or no. If desired, write out the gap.

Step 4. Develop countermeasures to close the gap. These countermeasures or actions should be specific, clear, objective, and relevant. Put them into a Gantt chart or an action plan.

Step 5. Execute/implement the plan.

If the decision is made to go for the future state, then implement the countermeasures on your action plan. Monitor and take actions if they do not remain on track.

The gap analysis tools used are:

- Interviews
- Standards
- Procedures
- Energy reduction checklist
- Critical success factors
- Review of central documents file that includes agendas, minutes, objectives, etc.

For interviews:

- Decide on who to interview.
- Will they be formal interviews or some random?
- Interview from 15-20 individuals for each self inspection.
- Develop a list of questions for the formal interviews and also for the random interviews.

Typical questions to ask management are:

1.   How is the energy management program going?

2.   Are you committed to the program as it is being implemented, or would you change something? If you would change something, please explain what it is.

3.   Have you received any energy management training?

4.   Do you think the corporate energy reduction goal is achievable and realistic?

5.   Would you support additional training for the organization's personnel?

6.   What additional training do you think would be helpful?

7.   Are you kept informed in a timely manner of energy management progress, problems, issues, and results?

8.   Do you mention the energy management goal, objectives, progress and results at your staff meetings?

Some typical questions to ask the people who you randomly pick as you are walking in the facility are:

1. Do you know two members of the organization's energy team?

2. Do you know who the energy champion is for the organization?

3. Name two ways you support the energy reduction program.

4. Did you receive energy awareness training?

5. Have you read the organization's energy policy?

6. Does your computer shut down when it has been idle for 30 minutes or more?

7. Are you comfortable with the temperature in your job location? Winter? Summer?

8. Have you received any energy conservation training?

9. Have you been on any energy team? Would you like to be a member if asked?

10. Have you noticed any changes that will save energy? If yes, what?

11. What items in your workstation use electricity?

In conducting the procedures reviews, the guidelines are:

1. What energy procedures does the organization have?

2. Are they current?

3. Are they readily available to everyone?

4. Is the subject adequately covered?

5. Are there missing areas not covered by a procedure that should be?

6. Are they numbered?

7. Are they reviewed for update at least annually?

EXCELLENCE FACTORS EVALUATION

The second method is called excellence factor evaluation. The method is:
- Comparison of current situation with excellence factors
- Excellence factors derived from energy management musts

First, an organization needs to write what their "musts" or mandatory requirements are for their strategic energy management system are. These musts will vary from organization to organization. Typical musts are listed below:

1. All organizations reducing energy must have an energy policy, an energy plan, and committed leadership.

2. All organizations must have an energy champion and a cross-functional energy team to be successful.

3. Top management, energy champion and energy team must know their roles and responsibilities.

4. All energy team members must understand electricity bills and costs incurred for such things as power factor.

5. All personnel must receive energy awareness training.

6. Energy performance indicators (EnPIs) must be graphed and displayed for everyone to see at the recommended frequency.

7. All EnPIs must have a data collection plan.

8. All objectives and targets must have an action plan.

9. All organizations must develop an energy conservation plan and present to all the organization's people.

10. A list of low hanging fruit should be developed and then implemented.

11. All energy projects must have a simple payback calculation.

12. Sufficient contribution must be identified to meet the corporate goal.

13. IT power management and reducing office paper objectives should be accomplished by all organizations desiring to reduce energy.

14. Data centers must not be cold unless energy projects to correct this situation have been developed.

15. Management reviews must be held at least annually.

16. An energy awareness culture should be built with training, communications, involvement, and recognition.

17. Perform periodically self inspections or internal audits.

Next, the excellence factors are identified that fully cover all of the must requirements. In the musts identified above, the excellence factors shown in Figure 8-1.

In Figure 8-2 Excellence Factors Evaluation, the self-inspection method cycle is shown.

The leaders should talk the talk (give speeches at staff meetings, town meetings and discuss energy reduction with management, employees and contractors. They walk the walk by practicing energy conservation themselves. They become the talk when everyone sees how committed they are, that this is their vision, and all need to support it. The organization must be focused on achieving energy cost savings and energy efficiency. To be focused, the organization needs management to complete a compelling energy policy, a 3- to 5-year, energy plan,

Figure 8-1. Excellence Factors List

#1 Leadership
#2 Focus
#3 Processes
#4 Communications
#5 Measurement
#6 Organization
#7 Management
#8 Culture
#9 Plans
#10 Contributions
#11 O&Ts
#12 Training
#13 Continuous improvement
#14 Self inspections

Figure 8-2. Excellence Factors Evaluation

a cross-functional energy team that establishes meaningful objectives and targets and manages energy reduction implementation. An energy deployment system such as Energy Star, ISO 50001 EnMS, DOE's Superior Energy Performance, or Energy Centered Management System covered in this book should be followed. Figure 8-4 shows these main elements of being focused.

In Figure 8-5, processes are important. They must be understood by the process participants, be efficient and be effective if energy reduction is to occur.

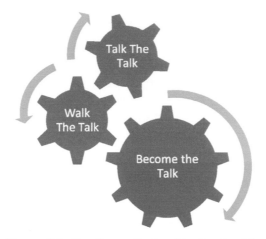

Figure 8-3. Excellence Factor #1—Leadership

FOCUS
Energy Policy/Training
Energy Plan
Energy Team Meetings
Energy Deployment Systems

Figure 8-4. Excellence Factor #2—Focus

Process Performance	• Outputs • Outcomes
Process Improvements	• Lean Techniques • Six Sigma
Process Additions	• Identifying Energy Waste • Problem Solving

Figure 8-5. Excellence Factor #3—Processes

Communication is extremely important if the energy reduction effort is to be achieved. Both internal and external communications are important. A communication plan covering both of these is recommended to be accomplished and explained to all the organization's personnel. The objective is to keep all personnel informed and energy aware so that the desired organizational culture can occur.

If you cannot measure progress and results, then there is no way to tell how things are going, and any road could be taken.

Having some organization established to do the planning, development and manage the implementation of energy management and reduction is a must. For smaller organizations, an energy manager will suffice. In larger organizations, an energy champion, a cross-functional energy team, a walkthrough team, facility managers at each facility, along with an energy engineer or adviser may be required to organize and manage the numerous activities and responsibilities of reducing energy consumption and cost.

Management is key to any organization to be run efficiently and effectively. The functions that support the energy management system such as human resources, resource management, and budgeting must include supporting the energy management system.

An energy-aware culture must be created to ensure the energy program's immediate success and to sustain the desired effort. Everyone must be energy aware, practice energy conservation, and identify energy waste and help minimize or eliminate it.

COMMUNICATIONS
External
Recognition and Emphasis
Internal
Progress & Results

Figure 8-6. Excellence Factor #4—Communications

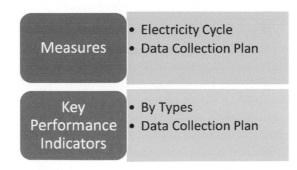

Figure 8-7. Excellence Factor #5—Measurements

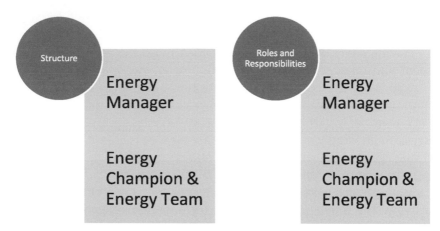

Figure 8-8. Excellence Factor #6—Organization

---

1. Resource Management
2. Project Management
3. Planning
4. Implementing
5. Budgeting

Figure 8-9. Excellence Factor #7—Management

CULTURE
Attitudes & Morale
Behavior
Values & Principles
Support & Involvement

Figure 8-10. Excellence Factor #8—Culture

No improvement program can be achieved without plans developed and then followed. Energy management cannot function efficiently and effectively without plans—communication, documentation, measurement, monitoring, and emergency to name a few. Universities and hospitals have many other plans to cover their energy management and related functions. Space assignments and scheduling, lab plans, and other plans are paramount in managing energy management at universities and colleges. Some key university plans are:

- Building automation master plan.

- Equipment energy awareness plan.

- Energy Star procurement policy & plan.

- Electrical demand response and management plan & program.

- Energy outreach plan.

- Long-term metering plan.

- Sustainability strategic plan.

- Campus climate action plan.

- IT energy plan.

- Building energy conservation incentive plan.

- Fume hood modernization program.

The self inspection needs to be sure these plans are meaningful, being used, current and complete. They must be:

- Current?   ___ Yes ___ No.

- Complete?   ___ Yes ___ No.

- Have a purpose?   ___ Yes ___ No.

- Responsibilities outlined?   ___ Yes ___ No.

- What and when shown for activities required or planned?

                                                                ___Yes ___ No.

- How or method of accomplishment? _____

- If requireing a team or committee, are they meeting regularly and keeping minutes?   ___ Yes ___ No.

PLANS
Communications Plan
Monitoring and Measurement Plan
Emergency Plan
Other Plans

Figure 8-11. Excellence Factor #9—Plans

Significant contributions to meet the corporate goal must be identified and estimated, and the percentage each will contributed once
implemented or completed to achieving the goal. These contributions
need to be tracked, monitored and re-evaluated when needed but at
least annually.

CONTRIBUTIONS

Objectives & Targets

Energy Projects

Energy-centered Maintenance

Low Hanging Fruit

Figure 8-12. Excellence Factor #10—Contributions

Objectives and targets with well developed action plans are the
drivers of the energy management program to achieve meaningful
milestones and maintain the desired program momentum.

Meaningful?

Responsible Person

Action Plan

Measure of Success

Number Adequate?

**Figure 8-13. Excellence Factor #11—
Objectives, Targets & Action Plans**

Energy awareness training and energy conservation training
should be given at least once a year to achieve the desired cultural,
support, and program sustainment. Other specific training should be
conducted once the requirement has been identified and verified.

The energy management effort is a journey with no destination.
Energy management will be required as long as organizations use
energy to achieve their mission and comfort of their employees

To sustain any program, self inspections or internal audits are
needed to determine what needs to be done to improve and assure
compliance with the requirements.

The excellent factors add an extra touch to this self inspection
method. The self inspection analysis can be recorded in this format that

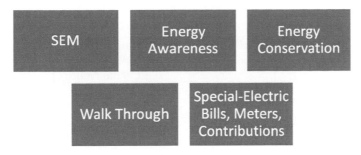

Figure 8-14. Excellence Factor #12—Training

Plan-Do-Check-Act
Sheward Cycle-Deming Wheel-
Continuous Improvement Wheel

Figure 8-15. Excellence Factor #13—
Continuous Improvement

Figure 8-16. Excellence Factor #14—Self Inspections

shows compliance with the musts and excellence factors and remarks that explain the deficiencies.

Filling out Table 8-1 by reviewing meeting minutes of energy meetings and management reviews, procedures and conducting interviews will give a strong indication of how well the energy management system is being implemented, the progress and any deficiencies or non-conformances to the must list.

Table 8-1

Musts Requirements	Excellence Factors	Compliance Full Almost Partial None	Remarks
All organizations reducing energy must have an energy policy and an energy plan and committed leadership.	Leadership`		
All Organizations Must Have an Energy Champion and a Cross-Functional Energy Team to Be Successful	Organization		
Top Management, Energy Champion and Energy Team Must Know Their Roles and Responsibilities.	Organization/ Management		
All Energy Team Members Must Understand Electricity Bills and Cost Incurred for Such Things as Power Factor.	Training/Processes		
All Personnel Must Receive Energy Awareness Training.	Training/ Communications		
Energy Performance Indicators (EnPIs) Must Be Graphed and Displayed for everyone To SEE at the recommended frequency.	Measurement/ Communications		
All EnPIs Must Have a Data Collection Plan.	Measurement		

Table 8-1. (*Cont'd*)

All Objectives and Targets Must Have An Action Plan.	Objectives and Targets		
All Organizations Must Develop an Energy Conservation Plan and Present to all The Organization's People.	Training/Plans/ Communications		
A List of Low Hanging Fruit Should Be Developed and Then Implemented.	Contributions/Plans		
All Energy Projects Must Have a Simple Pay Back Calculation.	Contributions/ Processes		
Sufficient Contribution Must Be Identified to Meet The Corporate Goal.	Contributions/ Management		
IT Power Management and Reducing Office Paper Objectives Should be Accomplished By All Organizations Desiring to Reduce Energy.	Contributions/ Management		
Data Centers Must Not Be Cold Unless Energy Projects To Correct This Situation Have Been Developed.	Contributions		
Management Reviews Must Be Held at Least Annually.	Organization/Focus/ Continuous Improvement		

Table 8-1. (*Cont'd*)

An Energy Awareness Culture should be build with training, communications, involvement, and recognition.	Culture		
Perform Periodically Self Inspections or Internal Audits.	Self Inspections /Continuous Improvement		

SELF INSPECTION/INTERNAL AUDITS CHECKLIST

Energy management programs have many areas that are expected to be performed efficiently and effectively to ensure desired results. This book will provide numerous assessment tools that when used will show the current status of your program. The expectations minus the current situation gives a gap showing the areas needing fixing. The gaps will need countermeasures which in turn need action plans or corrective actions reports. When all of the gaps are closed by completing the countermeasures, your energy management program will be healthy and thriving with positive results.

The audit checklist can be used for energy team self inspections or internal audits done by an audit team. The audit checklist has 12 major areas. They are:

- Organization and roles and responsibilities

- Energy policy

- Energy plan

- Objectives and targets and energy action plans

- Energy training

- Communications

- Monitoring and measurement

- Documentation

- Auditing and corrective action

- Management reviews

- Energy reduction deployment

The process is to develop what is expected and then use the audit checklist to identify the current situation. Then the formula Expectations – Current Situation = Gap Plans applies.

Then the gaps, countermeasures, actions plans or CARs will get the implementation back on track.

Expectations will consist of mandatory requirements and desired actions that have proven beneficial in the past. A gap is the difference between where you are and where you want to be.

The self-inspection checklist enables an internal audit of actual achievements versus standards requirements in 12 major areas. They are roles and responsibilities, energy policy, energy plan, objectives and targets and their actions plans, training, communications, documentation, monitoring and measurement, facility auditing and corrective actions, energy reduction deployment process and management reviews. This same checklist can be used by an organization's headquarters audit team in performing an audit of a facility or an overall organizational audit. The internal audit can be done by the headquarters internal audit department provided they develop an audit plan and assign auditors that have had some energy management training. This second-party audit needs to be accomplished every three years. The audit will consist using the self-inspection checklist, interviewing personnel from top management, employees to determine their awareness, the energy champion, the energy team members, and the facilities' managers and whomever else they desire to interview. They will confirm that they accomplished what the energy reduction deployment process implementation plan said they were going to do. A check on the energy performance indicators will show what progress and results have been achieved.

Self Inspections/Internal Audits

Roles and responsibilities expectations are listed below:

Expectation 1: Energy team member roles should be identified and documented in team meeting minutes and reported to the energy champion.

Expectation 2:   Team leader, facilitator, and note keeper should be appointed or selected.

Expectation 3:   Team members with identified roles should understand their responsibilities and perform them.

Expectation 4:   The energy team members should actively participate in meetings?

Expectation 5:   Facility staff and management should be aware of their responsibilities with regard to energy reduction deployment.

1.   Top management, energy champion and energy team must know their roles and responsibilities.   ___Yes ___No

2.   All organizations must have an energy champion or energy manager and a cross-functional energy team to be successful.
                                                                    ___Yes ___No

ELEMENT 1: Roles and Responsibilities

1.   Have energy team member roles been identified and documented in team meeting minutes and reported to the energy champion?
                                                                    ___Yes ___No

2.   Has a team leader, facilitator, and note keeper been appointed?
                                                                    ___Yes ___No

3.   Do team members with identified roles understand their responsibilities and are they being fulfilled?   ___Yes ___No

4.   Do the energy team members participate in meetings?
                                                                    ___Yes ___No

5.   Are facility staff and management aware of their responsibilities with regard to energy reduction deployment?   ___Yes ___No

Comments:

_____
_____
_____

Gaps

1. Roles and responsibilities identified and understand. _____

2. Team members, team leader, note taker, document control manager appointed. _____

3. Appointees are performing their duties and meeting their responsibilities. _____

4. High participation by team members at meetings. _____

5. Staff and management understand energy reduction program. _____

6. Top management, energy champion and energy team know their roles and responsibilities. _____

Energy Policy

Expectations 1:   Top management approved, and communicated to all, a compelling energy policy.

Expectations 2:   Energy policy is reviewed for possible changes each year by top management.

Mandatory Requirement: All organizations reducing energy must have an energy policy and an energy plan.   ___Yes ___No

ELEMENT 2: Energy Policy

1. Does the organization have an energy policy?   ___Yes ___No

2. Has the energy policy been approved by the senior leadership.
    ___Yes ___No

3. Is the policy current and reviewed annually?
    (Date of last review:_____)   ___Yes ___No

4. Has the policy been provided and communicated to management, employees and contractors?   ___Yes ___No

Comments:
_____
_____
_____

Gaps

1.   Have compelling energy policy.   ___Yes ___No
2.   Communicated energy policy to all personnel.   ___Yes ___No
3.   Is a procedure in place requiring periodic updates?

                                                      ___Yes ___No

Energy Plan

Expectations 1: A 3- to 5-year energy plan be developed and approved
                 by top management.
Expectations 2: The plan is reviewed annually for possible revisions.
Expectations 3: The plan includes the corporate energy goal and en-
                 ergy policy.

1.   All organizations reducing energy must have an energy policy
      and an energy plan.   ___Yes___No

ELEMENT 3: Energy Plan

1.   Does the organization have an energy plan?   ___Yes___No

2.   Does it cover at least 3 years?   ___Yes___No

3.   Is there a procedure in place to ensure the plan is periodically
      updated and when a major revision is in order? ___Yes___No

4.   Is the energy policy and the corporate goal included in the energy
      plan?   ___Yes___No

Comments:

_____

_____

_____

Gaps

1.   Have an energy plan that has been approved by top management.

                                                      ___Yes___No

2.   Energy plan is for at least a 3-year time period.

                                                      ___Yes ___No

3.   Energy plan includes the corporate energy goal and energy poli-
      cy.   ___Yes___No

Objectives and Targets and Energy Actions Plans
Expectations

1. Energy team worked on at least one objective and target this year.

2. Objectives and targets are documented.

3. Actions plans have someone designated as responsible, and the actions outlined when done will achieve the targets.

4. The energy team meetings reviewed action plans at the meeting.

5. The status using stop light colors gets updated when the status changes.

6. All objectives and targets must have an action plan.

7. All organizations must develop an energy conservation plan and present to all the organization's people.

8. A list of low hanging fruit should be developed and then implemented.

9. All energy projects must have a simple payback calculation.

10. Sufficient contribution must be identified to meet the corporate goal.

11. IT power management and reducing office paper objectives should be accomplished by all organizations desiring to reduce energy.

12. Data centers must not be cold unless energy projects to correct this situation have been developed.

ELEMENT 4: Objectives & Targets and Energy Action Plans (EAP)

1. Has the energy team worked on or initiated at least one objective and target this year?   ___Yes___No

2. Have objectives and targets and action plans been documented?
                                                          ___Yes___No

3. Do objectives and targets address the respective facility's energy efficiencies?   ___Yes___No

4.   Has a documented action plan designating responsibility and detailing steps to completion been put in place for each objective, and written to describe how the team will successfully meet the target?   ___Yes___No

5.   Are objectives being reviewed for progress at the team meetings?
                                                              ___Yes___No

6.   Is the status for each action plan action reflected using a stop light symbol at each meeting? (Date of last review: _____)
                                                              ___Yes___No

Comments:
_____
_____
_____

Gaps
1. Energy team worked on at least one objective and target this year. _____

2. Objectives and targets are documented. _____

3. Actions plans have someone designated as responsible and the actions outlined when done will achieve the targets. _____

4. The energy team meetings reviewed action plans at the meeting. _____

5. The stop light colors get updated when the status changes. _____

6. All objectives and targets must have an action plan. _____

7. All organizations must develop an energy conservation plan and present to all the organization's people. _____

8. A list of low hanging fruit should be developed and then implemented. _____

9. All energy projects must have a simple payback calculation. _____

10. Sufficient contribution must be identified to meet the corporate goal. _____

11. IT power management and reducing office paper objectives should be accomplished by all organizations desiring to reduce energy. _____

12. Data centers must not be cold unless energy projects to correct this situation have been developed. _____

Training Expectations

1. All personnel received energy awareness training.

2. The training plan has been reviewed and updated during the last year.

3. All personnel have received energy conservation training.

4. Targeted personnel have received the energy reduction deployment training.

Mandatory Expectations

1. All energy team members must understand electricity bills and costs incurred for such things as power factor.

2. All personnel must receive energy awareness training.

ELEMENT 5: Training

1. Has energy awareness training been developed and presented to middle management, employees and contractors? ___Yes___No

2. Have the training needs been reviewed within the past year? (Date of last review:_____)
   ___Yes___No

3. Have employees received energy conservation training?
   ___Yes___No

4. Have management, energy champion, and energy team received energy reduction deployment training? ___Yes___No

Comments:

_____

_____

_____

Gaps

1.   All personnel received energy awareness training. _____

2.   The training plan has been reviewed and updated during the last year. _____

3.   All personnel have received energy conservation training. _____

4.   Targeted personnel have received the energy reduction deployment training. _____

5.   All energy team members must understand electricity bills and costs incurred for such things as power factor and demand. _____

Communications Expectations

1.   The energy champion and/or energy team keeps management and employees informed on relevant information.

2.   All personnel know whom to communicate energy concerns or ideas.

3.   Everyone knows the procedure for sending or receiving external information.

Mandatory Expectations

1.   Energy progress and results must be communicated periodically to top management and to all personnel.   ___Yes___No

ELEMENT 6: Communication

1.   Do the energy champion and energy team communicate relevant information to the facility middle management, employees and contractors?   ___Yes___No

2.   Do employees know to whom to communicate energy concerns? Is there a communications plan?   ___Yes___No

3.   Are external communications adequately addressed?
                                      ___Yes   ___No ___N/A

Comments:

_____

_____

_____

Gaps
1.   The energy champion and/or energy team keeps management and employees informed on relevant information. _____

2.   All personnel know whom to communicate energy concerns or ideas. _____

3.   Everyone knows the procedure for sending or receiving external information. _____

Energy Monitoring and Measurement Expectations
1.   Significant electricity users (SEUs) that require monitoring or measurement have been identified on the monitor and measure plan.

2.   The monitor or measure plan has been reviewed within the past year.

3.   The required monitoring and measuring is being conducted for the energy performance indicators (EnPIs).

4.   The monitoring/measurement relevant to the projects under construction is being accomplished.

Mandatory Expectations
1.   The required EnPIs must be selected for use.

2.   Energy performance indicators (EnPIs) must be graphed and displayed for everyone to see at the recommended frequency.

3.   All EnPIs must have a data collection plan.

ELEMENT 7: Monitoring and Measurement

1.   Have significant electricity users that require monitoring or measurement been identified on the monitor and measure plan?

___Yes___No

2.   Has the monitor or measure plan been reviewed within the past year? (Date of last review:_____)   ___Yes___No

3.   Is the required monitoring and measuring being conducted for the energy performance indicators?   ___Yes___No

4.   Is the monitoring/measurement relevant to the projects under construction?   ___Yes___No

Gaps

1.   The required EnPIs must be selected for use. _____

2.   Energy performance indicators (EnPIs) must be graphed and displayed for everyone to see at the recommended frequency. _____

3.   All EnPIs must have a data collection plan. _____

4.   Significant electricity users (SEUs) that require monitoring or measurement have been identified on the monitor and measure plan. _____

5.   The monitor or measure plan has been reviewed within the past year. _____

6.   The required monitoring and measuring is being conducted for the energy performance indicators (EnPIs). _____

7.   The monitoring/measurement relevant to the projects under construction is being accomplished. _____

Documentation Expectations

1.   Required documents filed in the organization's designated document file.

2.   Documents are filed in the correct folder.

3.   Documentation is current.

4.   Documents used by the energy team and energy champion are current.

5.   Energy documents provide direction and instructions to other related documents, and schedule.

Mandatory Expectations

Essential documentation for self inspections and to ensure only current documents are used by the energy champion and energy team must be accomplished.   ___Yes ___No

ELEMENT 8: Documentation and Control of Documents

1.   Is all required documentation filed on the facility's energy document control site?   ___Yes ___No

2.   Is all documentation filed in the correct folders according to the energy documentation guide procedure?

     ___Yes ___No

3.   Is all documentation on the document control site current?

     ___Yes ___No

4.   Are the documents used by the energy team the current energy program documents?   ___Yes ___No

5.   Does the energy documentation provide direction and instructions to other related documents, records, reports, schedules and registers?   ___Yes ___No

Comments:

_____

_____

_____

Gaps

1.   Required documents filed in the organization's designated document file. _____

2.   Documents are filed in the correct folder. _____

3.   Documentation is current. _____

4.   Documents used by the energy team and energy champion are current. _____

5.   Energy documents provide direction and instructions to other related documents, and schedule. _____

Auditing and Corrective Actions Expectations
1.   Self inspections by the energy team are conducted at least once a year.

2.   Non-conformances or discrepancies are identified and corrective actions taken.

3.   All corrective actions reports are completed and verified on schedule.

Mandatory Expectations
1.   A self inspection of the energy management program must be accomplished by the energy team at least annually.

___Yes ___No

ELEMENT 9: Facility Auditing and Corrective Action

1.   Have self inspections been completed annually? ___Yes ___No

2.   Were all non-conformances identified and corrective action taken to address the non-conformance?   ___YES ___NO ___N/A

3.   Were all CARs from the previous second-party audits resolved appropriately?   ___YES ___NO ___N/A

Comments:
_____
_____
_____

Gaps
1.   Self inspections by the energy team are conducted at least once a year. _____

2.   Non-conformances or discrepancies are identified and corrective actions taken. _____

3. All corrective actions reports are completed and verified on schedule. _____

Management Reviews Expectations

1. Management review are conducted at least once a year.

2. The head of the facility is up to speed on the status of the energy reduction program.

3. The head of the facility provides guidance and direction to the energy reduction program.

4. The required inputs and outputs of a management review are followed and accomplished.

Mandatory Expectations

1. Management reviews must be held at least annually.

_____Yes _____No

ELEMENT 10: Management Review

1. Has the management review been conducted at least annually? (Describe the management review process and date(s) of last review: _____)

_____Yes _____No

2. Is the head of the facility, or their designee, aware of the status of the energy reduction program? _____Yes _____No

3. Does the head of the facility, or their designee, provide guidance and direction for the energy reduction deployment?

_____Yes _____No

4. Does the management review cover the appropriate elements including both inputs and outputs? _____Yes _____No

Comments:

_____

_____

_____

Gaps
1.   Management review are conducted at least once a year._____
2.   The head of the facility is up to speed on the status of the energy reduction program._____
3.   The head of the facility provides guidance and direction to the energy reduction program._____
4.   The required inputs and outputs of a management review are followed and accomplished. _____

Energy Reduction Expectations
1.   Top management, energy champion, and energy team receive one of these training sessions:

    —   ISO 50001 Energy Management System.
    —   Energy centered management system.
    —   Organization's energy management strategic energy plan.
2.   Walkthroughs identified energy waste in the facility for all categories.
3.   The ECM has been put into the CMMS.
4.   The ECM has been scheduled for accomplishment.
5.   Good energy projects been identified for future funding.
6.   Energy team has developed sufficient contribution to meet corporate goal.

Mandatory Expectations
1.   A deployment system must be used to ensure all important areas are covered in the energy management program or system.
                                                    ___Yes ___No

ELEMENT 11: Energy Reduction Deployment
1.   Were the walkthroughs successful in identifying the energy information necessary for ECP and ECM?   ___Yes ___No

2.   Has the ECM been placed into a CMMS?   ___Yes ___No

3.   Has CMMS scheduled and accomplished the new preventative maintenance items?   ___Yes ___No

4.   Have good projects been identified to reduce energy use?

                                                    ___Yes ___No

5. Has the energy team identified significant contribution to achieve the corporate objective? ___Yes ___No

Comments

_____

_____

_____

Gaps

1. Top management, energy champion, and energy team receive one of these training sessions:
    *   ISO 50001 Energy Management System
    *   Energy centered management system
    *   Organization's energy management strategic energy plan
    *   Energy Star energy management system

1. A deployment system must be used to ensure all important areas are covered in the energy management program or system.

                                            ___Yes ___No

2. Walkthroughs identified energy waste in the facility for all categories. _____

3. The ECM has been put into the CMMS. _____

4. The ECM has been scheduled for accomplishment. _____

5. Good energy projects been identified for future funding. _____

6. Energy team has developed sufficient contribution to meet corporate goal. _____

Plans Expectations (underlined are mandatory)
*   Emergency plan
*   Communications plan
*   Procurement plan
*   Documentation plan
*   Monitoring and measurement plan
*   Energy plan
*   Training plan
*   All reviewed yearly for possible revisions.

ELEMENT 12: Plans

1. Does the organization have an emergency plan? ___Yes ___No

2. Does the organization have a communications plan?

                                                        ___Yes ___No

3. Does the organization have a procurement plan?___Yes ___No

4. Does the organization have a documentation plan?

                                                        ___Yes ___No

5. Does the organization have a monitoring and measurement plan?

                                                        ___Yes ___No

6. Does the organization have an energy plan?   ___Yes ___No

7. Does the organization have a training plan?   ___Yes ___No

8. Are all the plans reviewed annually by the energy team to see if they need revising?   ___Yes ___No

9. Does the organization have a contingency plan? ___Yes ——No

Gaps

- Emergency plan _____

- Communications plan _____

- Procurement plan _____

- Documentation plan _____

- Monitoring and measurement plan _____

- Energy plan _____

- Training plan _____

- Contingency plan _____

CORRECTIVE ACTIONS REPORTS (CARs)

CARs are used to solve deficiencies or non-conformances. Each is given a number and tracked until the problem has been eliminated and verified that it was. There are four parts to each CAR.

Table 8-2. Gaps—Countermeasures

Gap (Only if You Checked as a Gap)	Countermeasure
1. Roles and Responsibilities Identified and Understand.	Teach The Energy Champion, Energy Team and Walkthrough Team their Roles and Responsibilities shown previously.
2. Team Members, Team Leader, Note take, Document Control Manager Appointed	Top Management or Energy Champion should appoint.
3. Appointees are performing their duties and meeting their responsibilities	Will be determined by answers to later gap answers.
4. High Participation by Team Members at their Meetings	Have Team put attendance in minutes or use sign in sheets. See TOTs shown later in Countermeasures Area.
5. Staff and Management Understand Energy Reduction Program	Provide ECMS Training.
6. Top Management, Energy Champion and Energy Team Know Their Roles and Responsibilities.	Interview them and if they do not, teach them.
7. Have Compelling Energy Policy.	Develop using guidance in this course.
8. Communicated Energy Policy to all Personnel.	If it has not been done, send it by email and post on bulletin boards.
9. Have a Procedure in Place Requiring Periodic Updates of the Energy Policy.	Write an organization procedure and get top management's approval.
10. Have Energy Plan that has been approved by top management.	Develop using table of contents included in this course.
11. Energy Plan is for at Least Three Years Time Period.	If not, extend to three years by showing objectives, actions, and/or projects by year for three years.
12. Energy Plan Includes the Corporate Energy Goal and Energy Policy.	If not, include in an update.
13. Energy Team worked on at least one Objective and Target this year	If not, establish at least one objective with an energy action plan.
14. Objectives and Targets are Documented	Use organization's main operating system or SharePoint and set up a Documentation Control System per instructions included in this course..
15. Actions Plans have Someone Designated as Responsible and the Actions Outlined when Done will Achieve the Targets	If not, Start with current O & Ts and continue practice in the future. Ensure Energy action plans tell who is going to do what when.
16. The Energy Team Meetings Reviewed Action Plans at the Meeting	If not, add to the meeting agenda foreach meeting.
17. The Status Using Stop Light Colors Get Updated when the Status Changes.	Use Stop Light Status when appropriate.
18. All Objectives and Targets Must Have An Action Plan.	If not, do. Without an action plan, nothing will be accomplished.
19. All Organizations Must Develop an Energy Conservation Plan and Present to all The Organization's People.	Develop on power point so it can be emailed or used to teach in classroom.
20. A List of Low Hanging Fruit Should Be Developed and Then Implemented	Use the list shown later.

Table 8-2. Gaps—Countermeasures (*Cont'd*)

Gaps (Only if checked)	Countermeasures
21. All Energy Projects Must Have a Simple Pay Back Calculation.	See Later in Countermeasures Continued.
22. Sufficient Contribution Must Be Identified to Meet The Corporate Goal.	See Later in Countermeasures Continued.
23. IT Power Management and Reducing Office Paper Objectives Should be Accomplished By All Organizations Desiring to Reduce Energy	Use slides in this course and develop an O & T & Action Plan
24. Data Centers Must Not Be Cold Unless Energy Projects To Correct This Situation Have Been Developed.	Use slides in this course and develop an O & T & Action Plan.
25. All Personnel Received Energy Awareness Training	If not, develop an energy awareness training using guidance in this course.
26. The Training Plan has been Reviewed and Updated during the last Year.	Developed a Training Plan. See Later in Countermeasures Continued.
27. All Personnel have Received Energy Conservation Training	Develop Energy Conservation Training on power point and email to everyone in your organization.
28. Targeted Personnel have Received the Energy Reduction Deployment Training	Show them ECMS or an ISO 50001 EnMS Presentation.
29. All Energy Team Members Must Understand Electricity Bills and Cost Incurred for Such Things as Power Factor and Demand.	Provide the training in this presentation. Have copies of your electric bills ready for them to review.
30. The Energy Champion and/or Energy Team Keeps Management and Employees Informed on Relevant Information	Should communicate by EnPIs or tidbits about progress and results.
31. All Personnel Know Whom To Communicate Energy Concerns or Ideas.	Write a Communications plan or procedure and have this info as part of it.
32. Everyone Knows the Procedure for Sending or Receiving External Information	Write a Communications plan or procedure and have this info as part of it.
33. The Required EnPIs Must Be Selected for Use	Select KWH and CF Per Month, Electricity Intensity and Natural Gas Intensity, Electric Load Factor, Cost per Month and Costs/Sq. Ft., & % Renewable Energy
34. Energy Performance Indicators (EnPIs) Must Be Graphed and Displayed for Everyone To SEE at the Recommended Frequency.	Use excel
35. All EnPIs Must Have a Data Collection Plan.	Develop a Data Collection Plan like in the Countermeasures area.
36. Significant electricity users (SEUs) that require monitoring or measurement have been identified on the *Monitor and Measure Plan*	Identify the SEUs and place the ones requiring monitoring on the Monitoring and Measurement Plan.
37. The *Monitor or Measure Plan* has been reviewed within the past year.	Ensure Energy Team Reviews at lest once a year.
38. The required Monitoring and Measuring is being conducted for the Energy Performance Indicators (EnPIs)	The EnPIs are listed on the plan.

Table 8-2. Gaps—Countermeasures (*Cont'd*)

Gaps (Only if Checked)	Countermeasures
39. The Monitoring/Measurement relevant to the Projects under Construction is being Accomplished	List energy projects under construction on the M&M Plan.
40. Required Documents Filed In the Organization's Designated Document File	See Document List in the countermeasures section.
41. Documents are Filed in the Correct Folder	Randomly check about 30 Items every six months to ensure accuracy.
42. Documentation is Current	Randomly check about 30 Items every six months to ensure currency.
43. Documents Used by The Energy Team and Energy Champion are Current	Check the document numbers to ensure they are the latest ones at several Energy Team Meetings.
44. Energy Documents Provide Direction and Instructions to Other related Documents, and Schedule.	Keep accurate and current and use.
45. Self Inspections by the Energy Team are Conducted at least Once a Year	Check Document Control System to see if the self inspections are documented by year. If not, do so.
46. Non-Conformances or Discrepancies are Identified and Corrective Actions Taken	Check for CARs and whether they are closed out or not
47. All Corrective Actions Reports are Completed and Verified on Schedule	Check The CARs.
48. Management Review are Conducted at least Once a Year	Check Document Control System to see if the Management Reviews are documented by year. If not, do so.
49. The Head of the Facility is up to speed on the Status of the Energy Reduction Program	Conduct interviews during self inspection.
50. The Head of the Facility Provides Guidance and Direction to the Energy Reduction Program	Check with Energy Team and if not recommend to the appropriate person to do so.
51. The Required Inputs and Outputs of a Management Review are Followed and Accomplished	If not, do so. Include on the Management review's agenda.
52. A Deployment System must be used to ensure all important areas are covered in the Energy Management Program or System.	Use ECMS.
53. Walkthroughs Identified Energy Waste in the Facility for All Categories.	Add energy waste to the Energy Waste List.
54. The ECM Has Been Put Into the CMMS	Check CMMS. If not, do so.
The ECM Has Been Scheduled for Accomplishment	Check CMMS Reports and if not, do so.
55. Good Energy Projects Been Identified for Future Funding by Energy Team	Check to see if Energy Projects with Paybacks five years and under have been placed on an ECM.

Table 8-2. Gaps—Countermeasures (*Cont'd*)

Gaps (Only if checked)	Countermeasures
56. Energy Team Has Developed Sufficient Contribution to Meet Corporate Goal	Contribution should add up to the amount to achieve the goal. If not done, use technique included in this course.
57. Have an Emergency Plan	Develop IAW this Course.
58. Have a Communications Plan	Develop one that shows how internal and external computation should be handled.
59. Have a Procurement Plan	Develop for appliances , equipment and electronics. Use Energy Star and EPEAT.
60. Have a Documentation Plan/Procedure.	Show what has to be filed, who is responsible, frequency, and document and Record disposition.
61. Have a Training Plan	By Major Categories such as Top Management, Energy Team, Employees, Contractors what training they need and when they should receive it. Document the training on their training records.
62. Have a Contingency Plan	Include how energy availability will be ensured.

Four Parts

1. Nonconformance description
2. Cause analysis
3. Corrective action
4. Acceptance

Description of nonconforming work or departure from policies and procedures in the energy management system (EnMS) or energy regulations:
Car #:
ECAR #1
Policy reference:
Self-inspections audit checklist questions
Date issued:

Requirement:

Nonconformance:

Question #s, on EnMS audit checklist have not been done satisfactorily

Corrective action issued by:

Corrective action assigned to:
EMS team chief,
Correction action due by:

Figure 8-17. CAR Report

CAUSE ANALYSIS
EnMS team meeting date (if applicable):

Root cause analysis:

CORRECTIVE ACTION
Response:

Supporting documentation:

ACCEPTANCE
Additional audits required?   ___Yes ___No

Corrective action accepted?   ___Yes ___No

Associated preventive action request (PAR) (if applicable): _____

_____ Accepted:

EnMS team leader signature ——————————————
_____ Date accepted

_____

CARs are a formal, structured approach to solving problems. After completion, they should be documented for self inspections and or internal or external audits. The CARS should also be one of the inputs to the management reviews.

Chapter 9

Creating an Energy Reduction Culture and Emphasizing Energy Conservation

DEVELOPING AN ENERGY REDUCTION CULTURE

Organizational Cultural Factors

Creating a desired organizational culture sounds good, but what does it take to actually change an organization's cultural? Can management alone change the culture? How do we know if it is changed? The organizational culture that we would like to create is one where everyone is focused on reducing energy consumption. In other words, the organization's personnel identify energy waste in their work area and anywhere else in the facility they may see it, and they practice energy conservation daily. The factors that can help change an organization's culture are shown in Figure 9-1.

Figure 9-1. Organizational Culture

Organizational Cultural Impact Factors

- **Leadership**—A management commitment is a must for success. Leaders are better when they have a vision or a policy.

- **Vision, Policy, and Goal**—
  — Vision—Reaching a state of betterment.
  — Policy—A commitment from management as to what they or the employees are going to do.
  — Goal—To improve something.

- **Values, Principles and a Standard**—
  — Management desires for all personnel to possess and follow on a daily basis.
  — Examples: Committed to excellence, integrity, service before self, respect for people, safety first.

- **Attitudes and Morale**—What managers and employees believe and express to others. The morale is determined by whether the personnel are positive or negative or somewhere in between.

- **Communications**—Uses of several different media to inform, make aware or teach organization's managers, employees, and contractors.

- **Roles, Responsibilities, and Measurement**—
  — Key players know what is expected of them and their authority. Have measures that show progress and results.

- **Business Processes**—The way an organization gets its work done, achieves its mission, and meets its customers' needs.

These are our impact factors. Some or all must be changed to obtain a different culture than the present one.

- Leadership
- Vision, policy, goals
- Values, principles and standard
- Attitudes and morale
- Communications
- Roles, responsibilities and measurement
- Business processes

The above factors can impact an organization's culture depending on the emphasis and current situation. Leaders giving speeches at any opportunity can build the energy awareness culture needed to sustain energy management communications, the energy policy, the energy metrics and success stories. Let's find out your organization's current status.

Assessing an Organizational Culture

A survey questionnaire is shown below to assess your organization's culture. It is good to let the energy team and energy champion take this questionnaire after they are appointed and functioning. Take again after 6 months or a year and note what positive changes, if any, have occurred.

Organizational Cultural Assessment

> **Top Management**
>   1. ___ Considered to be like parents—promotes honesty and openness.

>   2. ___ Considered to be pacesetters—hard drivers and completely in charge.

>   3. ___ Once our leaders decide on a course of action, it will get done.

>   4. ___ They know what they want when we show them.

> **Strategic Planning and Emphasis**
>   1. ___ Does your organization have a strategic plan with a vision and corporate goals?

>   2. ___ Is there a corporate goal to reduce energy cost and usage?

>   3. ___ Is there an energy policy? If not, is there an environmental policy, quality policy or safety policy? Does the vision include reducing energy or operating cost or productivity?

>   4. ___ Do you have a strategic council, leadership council or something similar in place?

> **ISO Standards**
>   1. ___ Have you implemented or are you considering imple-

menting ISO 50001 Energy Management System (EnMS)?

2. ___ Have you implemented or are you considering implementing ISO 9001 Quality Management System (QMS)?

3. ___ Have you implemented or are you considering implementing ISO 14001 Environmental Management System (EMS)?

4. ___ Have you implemented or are you considering implementing OHSMS 18000 Safety Management System (SMS)?

5. ___ Has the organization appointed any management representative to lead an ISO standard implementation effort, and has she or he defined the roles and responsibilities of those involved?

> **Organization Characteristics**
1. ___ People in the organization are close and committed to the organization.

2. ___ Our organization is innovative and on the cutting edge.

3. ___ Our organization's primary emphasis is on production and objectives attainment.

4. ___ Our organization's heart is policies, procedures, and work rules.

> **Values**
1. ___ Our corporate values and principles are not known or not published.

2. ___ Our corporate values are published but not communicated well to all employees.

3. ___ Our corporate values are communicated to all our people and we believe in them.

4. ___ Our corporate values and guiding principles do not make sense.

> **Communications**
1. ___ Our organization communicates very well and uses more than one media.

2. ___ Communications is not our organization's strong point.

3. ___ All important performance indicators including electricity kWh usage are kept current and posted where all can see.

4. ___ Management holds regular staff meetings and energy usage is sometimes discussed.

➢ **Climate and/or Morale**
1. ___ The attitudes, feelings, and perceptions of individuals here are very positive.

2. ___ The morale seems to change whenever a major event occurs such as a change in leadership.

3. ___ Most everyone is proud to be a part of our company.

4. ___ The company sucks. I have a job and that is about it.

➢ **Training/Conservation**
1. ___ Our company focuses on keeping its employees competent and up-to-date.

2. ___ We only have orientation training, and occasionally some the boss feels important.

3. ___ We have had some six sigma and lean training.

4. ___ We have had energy conservation training but do not practice it.

5. ___ We have had energy conservation training and actively participate in reducing costs.

6. ___ Power management has been implemented and we purchase only energy saving computers and equipment.

- For each area, check your score and add for a total score. How is your culture? What areas need improvement (if any)?

- Top Management Q1-5 Pts, Q2-5 Pts, Q3-7 Pts, Q4-0 Pts (17 points possible)

- Strategic Planning Q1-7 Pts, Q2-7 Pts, Q3-7 Pts, Q4-6 Pts. (27 Points Possible)

- ISO-Q 1-7, Q2-4, Q3-5, QA4-4, Q5-7 (27 points possible)

- Organizational Characteristics Q1-7, Q2-7, Q3-5, Q4-3 (22 points possible)

- Values Q1-0, Q2-2, Q3-7, Q4-1 (10 points possible)

- Communications Q1-7, Q2-0, Q3-7, Q4-6 (20 points possible)

- Climate & Morale Q1-7, Q2-5, Q3-7, Q4-0 (19 points possible)

- Training Q1-7, Q2-5, Q-3-5, Q4-2, Q5-7, Q6-7 (33 points possible)

- Roles and Responsibilities Q1-4, Q2-5, Q3-7,Q4-6 (22 points possible)

- Business Processes Q1-5, Q2-6, Q-3-7, Q4-7, Q5-7 (32 points possible)

- Add up your score and see where your organization is toward creating an energy reduction culture.

Scores	Ease of Implementation	Probability Of Success	Comments
200+	Excellent	Excellent	
180+	Very Good	Very Good	
165+	Good	Good	
150+	Good	Good	Top Management support will significantly increase ease of implementation and probability of success
130+	Fair	Fair	
Below 130	Not So Good	Low	

Figure 9-2. Ease of Implementation and Probability of Success

DEVELOPING AND IMPLEMENTING AN
ENERGY CONSERVATION PROGRAM

If the organization has an energy manager or an energy team, development and implementation of an energy conservation program should be their responsibility. The intent is to get everyone involved in conserving energy. From an environmental viewpoint, a resource once consumed is not renewable, unless the energy used at the organization is renewable energy from solar, wind or geo-mass. Generating the electricity caused natural gasses to be released to the atmosphere. From a business viewpoint, energy costs money—money that can be used paying salaries or meeting other expenses.

In developing an energy conservation program, the energy manager or energy team should brainstorm a list of things that management, employees and on-site contractors can do to help conserve electricity. Better yet, go to the employees, provide them energy awareness training and then conduct a brainstorming session with the topic "What energy waste is occurring in your work areas now, or How can we save energy in your work area?" Be sure to coordinate the lists with facilities or engineering and get their cooperation and input. Conducting management/employee brainstorming sessions can generate excellent ideas. Also, it builds "buy-in" to the energy conservation program since they help create it.

Another option that has proven beneficial is to create several functional energy teams in large areas and have them identify ways to reduce waste in their area.

Once the energy conservation program is developed, it should be communicated by placing the requirement in an organization policy or procedure, placed on key posters and made visible throughout the facility, and a PowerPoint training program developed and sent to everyone as required training. Training sessions with a trainer or speeches at staff meetings and other meetings by management emphasizing the importance are also very beneficial.

GUIDING PRINCIPLES

It is helpful to develop a set of guiding principles for the program and use them in its development. An example is shown below:

1.   Energy manager and facilities conduct a monthly walkthrough.

2.   Start with low cost, high impact actions.

3.   Research promising technologies and areas to determine feasibility and appropriateness.

4.   Develop projects with high impact and short pay back period.

5.   Reduce kWh usage to meet organizational goals and/or company objectives.

6.   Measure progress and results and communicate to all.

ENERGY CONSERVATION BY AREAS OF IMPACT

As a rule, electricity is the primary energy element used. A good estimate would be 42% used for HVAC, 22% for lighting, 20% for office equipment, and 16% for other requirements such as hot water, special equipment, etc.

Implementing IT power management, reducing office paperwork, following a procurement policy for purchasing energy friendly appliance and electronics are energy centered measures. They belong here because it takes all of an organization's personnel to maintain and make them effective.

Walkthrough/FAME

The energy manager and facilities manager should periodically walk through the facility looking for equipment that is always running and excessive electrical items used for personal reasons such as microwaves, refrigerators, stereo systems, and TVs. Invite management in the areas of the walkthrough to join in the inspection. Check thermostat settings to ensure they are at the agreed upon winter (68 degrees) and summer (78 degrees) settings. If you do not, then establish one with management approval. One degree improvement above theses settings could reduce the electricity cost by 7% for the month.

•   Facilities will develop a FAME (facilities and management energy) checklist.

•   Visit each area once a month.

Figure 9-3. Energy Conservation Measures by Impact Area

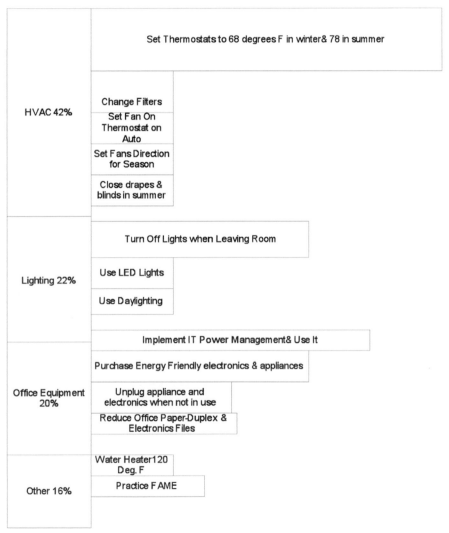

- Ensure thermostat settings are at agreed winter or summer settings.

Unplug Appliances When Not in Use

Management, employees and contractors are encouraged to turn off all electronics, appliances, radios, TVs, etc. when not in use, on weekends, and holidays. These items use electricity if they are plugged in even when they are not being used.

Limit Personal Items in Work Stations and Immediate Areas

Over time many appliances began to appear in work stations. Fans, heaters, small refrigerators, stereos, radios, televisions, and clocks. The list grows over time. All use electricity and run most of the time. Their being unplugged or heaters being disallowed should be a major focus of the FAME visits.

Turn-off-Lights Programs

This is an easy one to implement. Develop signs and place by conference room lights that remind, "Last one out, please turn off lights." Little stickers that say "turn off lights" can be purchased and used to signal the desired action. In areas where lights can be controlled as to on or off, designate someone to ensure they are turned off after a shift if no one is working in the area.

Facilities or engineering should note the areas that are infrequently visited such as break rooms, rest rooms, mechanical rooms, printing rooms, and supply rooms and initiate a program to install occupancy sensors in these areas. They will pay for themselves in less than two years. The same for occupancy sensors can be used in vending machines. In hallways and other areas where high light intensity is seldom required, install and use dimmers.

Establish Energy Procurement Policy

Have management establish a procurement program to buy only energy friendly appliances and electronics in all future purchases. Purchase Energy Star electronics and appliances when available to meet your requirements. Have a standard specification added to all bids ensuring energy friendly electronics, appliances, and motors are purchased.

Power Cords

Use power cords that turn off when not in use.

Duplexing/Electronic Files

Use double-sided copying as much as possible. Use electronic files as much as possible and avoid having both electronic and paper files.

IT Power Management

Implement IT power management for both your monitor and CPU. Allow for it to turn off after 15 minutes of being idle.

Use Day Lighting

Use day lighting when possible and turn down (dim) the lights.

Set Fans for Season

Fans should be set to go clockwise in winter and counterclockwise in summer.

Set Thermostats to Seasonal Recommended Settings and Have Fans on AUTO

Set thermostats to 68 degrees F in winter and 78 degrees F in summer. This will require employees to possibly wear some additional clothing and adjust to the new settings.

Change Filters

Changing heating and air conditioning filters on a regular schedule keeps HVAC from working harder than necessary plus enables a cleaner work environment.

Water heaters

Set water heaters to 120 degrees F and keep them at this setting.

LED Lights

Where possible change out lights and use LED lights instead.

If everyone participates in these items, possibly a five percent reduction in energy consumption and cost can be saved. Energy conservation must become a habit and always part of our behavior. Below is an example of an energy conservation poster. Posters should be placed throughout the work area emphasizing energy conservation. The energy conservation training given on PowerPoint slides as an email attachment should be done yearly.

ENERGY CONSERVATION

Practice energy conservation. Implement "Last one out, turn off lights policy." Turn off the lights when you leave the room.

1.   Turn off electrical fixtures not needed and unplug them if leaving for the day or weekend. Unplug items when not in use. Equipment uses electricity even when it is turned off.

2.   Don't use excessive personal items that require electricity. Fans, heaters, microwaves, refrigerators, etc., are high-energy users and are in violation of safety rules. To use them in your area, you need the building manager's approval in writing.

3.   Use power cords that turn off when not in use. Can be purchased from FEMP products and other stores that sell energy friendly products.

4.   Use duplex printing at every opportunity and electronic files.

5.   If working on a weekend, only turn on lights that are absolutely needed.

6.   Practice power management on your computers, printers, laptops, and imaging machines when possible and per instructions from IT.

7.   When ordering or purchasing electrical items, buy Energy Star, and for computers, monitors, and laptops, purchase EPEAT.

Figure 9-4. Energy Conservation Poster

Chapter 10

IT Power Management

WHAT IS POWER MANAGEMENT?

In today's environment, just about every white collar worker plus many blue collar workers use computers to accomplish their work tasks. A large organization could have thousands of computers and monitors running at any time. Computers and monitors use electricity and electricity costs money. Power management is a method of reducing electricity used on key office equipment by allowing them to enter a low power sleep mode when they are inactive. Also, it includes supervision, encouraging employees to turn off or unplug equipment when not in use or when they leave. The IT power management program is designed to have the computers and monitors go to a much less electricity consumption mode when they have not been busy for 20 or 30 minutes. This action can save $30-35 annually for each monitor and computer when implemented.

The IT power management program should include:

- Purchase Energy Star electronics equipment.

- Enable Energy star sleep function.

- Minimize screen savers.

- Reduced time from not in use until computer and monitor go into sleep mode.

- Ensure Energy Star use and turning equipment off when not in use.

PURCHASE ENERGY STAR COMPUTERS,
MONITORS, AND OTHER ELECTRONICS

ENERGY STAR is a government-backed program helping businesses and individuals protect the environment through superior energy efficiency (https://www.energystar.gov).

The ENERGY STAR label was established to:

- Reduce greenhouse gas emissions and other pollutants caused by the inefficient use of energy; and make it easy for consumers to identify and purchase energy-efficient products that offer savings on energy bills without sacrificing performance, features, and comfort.

Products can earn the ENERGY STAR label by meeting the energy efficiency requirements set forth in ENERGY STAR product specifications. EPA established these specifications based on the following set of key guiding principles:

- Product categories must contribute significant energy savings nationwide.

- Qualified products must deliver the features and performance demanded by consumers, in addition to increased energy efficiency.

If the qualified product costs more than a conventional, less-efficient counterpart, purchasers will recover their investment in increased energy efficiency through utility bill savings, within a reasonable period of time.

- Energy efficiency can be achieved through broadly available, non-proprietary technologies offered by more than one manufacturer.

- Product energy consumption and performance can be measured and verified with testing.

- Labeling would effectively differentiate products and be visible for purchasers.

- Develop and communicate a procedure and policy to purchase only energy saving electronics and appliances. Develop a standard requirement to be entered in the specifications for purchase of electronics and appliances.

ENABLE ENERGY STAR SLEEP FUNCTION

Office equipment that has earned the ENERGY STAR® helps eliminate wasted energy through special power management features. When equipment is not in use, it can be configured to automatically enter a low-power "sleep" mode. An ENERGY STAR qualified computer in sleep mode consumes about 80% less electricity than it does in full-power mode. Overall, ENERGY STAR qualified office products use about half as much electricity as standard equipment, and using less energy keeps utility costs down. In 2003, ENERGY STAR-qualified home office equipment saved Americans more than $3.5 billion in energy costs.

- Monitor Power Management (MPM): places active *monitors* into a low power sleep mode after a period of inactivity. *Can save $10 to $30 per monitor annually.*

- Computer Power Management (CPM): System standby and hibernate features place the *computer* itself (CPU, hard drive, etc.) into a low power sleep mode. *Can save $15 to $45 per desktop annually.*

- Good settings P.M. are to have the computer or laptop go to sleep (system standby or hibernate) after 30 minutes of inactivity and the monitors go to sleep after 15 minutes of inactivity.

- If this is not happening, check your settings or consult with IT.

MINIMIZE SCREEN SAVERS

Screen savers with modern computers are not a necessity as they once were. Most operating systems will support screen savers. They look good and many can be obtained free by just downloading them. Landscaping including beautiful scenery, pets, animated screens, holidays, oceans, and outer space are a few that can be available. Should we use them? Screen savers use energy and are not needed to protect the computer as they did for some of the old monitors.

❖Built into Windows 95,98, ME, 2000, XP , 7   and now Vista
❖Settings simply need to be activated

Figure 10-1. Power Management Features

REDUCE TIME OF IDLE

Organizations normally set a policy for the time that an employee may have the computer and monitor idle before it goes into sleep mode or hibernates. Some organizations use 30 minutes. Others start with 30 and then later reduce to 15 or 10 minutes to achieve further savings.

ENSURE ENERGY STAR USE AND TURNING EQUIPMENT OFF WHEN NOT IN USE

Recommend to procurement that they develop and get management approval requiring energy star purchase for any new electronics or appliances. This requirement should be put into an organization procedure and communicated to everyone.

Supervisors and management should encourage that computers and monitors be unplugged at end of shift and/or holidays or vacations. Check with IT first, since often they upgrade computers with patches and the computer and monitor must be on to receive the patch. There are several solutions to this latter problem with using either free software from EPA or purchasing software that monitors that power management is being used and can even turn computers and monitors on, upgrade them and then turn them off.

Chapter 11

Reducing Office Paper Use

OBJECTIVE

Our objective is to reduce office paper use so we can save energy. There are other objectives that go with this key objective. They are:

1. Buy paper with 30% recycled content.

2. Reduce amount purchased.

3. Use paper efficiently.

4. Recycle the used paper.

The objectives include the entire life cycle as shown in Figure 11-1.

PAPER USE LESS PROGRAM
(PULP)

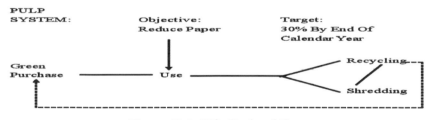

Figure 11-1. Life Cycle of Paper

Objective 1. Buy paper with 30% recycled content. All government organizations are required to purchase office paper with at least 30% recycled content. Many have experimented with higher content to ensure they work well in their imaging machines and printers. Some organizations are buying 100% recycled now, and most are purchasing paper greater than 30% recycled.

Objective 2. Reduce amount purchased. If the major objective is done, then this will be achieved also. Organizations have reduced paper use as much as 50% in two years when they have made it as a corporate objective. The more we save, the less we need to purchase. At the beginning of the year count the number of reams on hand, then count at the end of year. Amount used for the year = amount reams on hand Jan 1 plus reams purchased during the year – the amount on hand at the end of the year. (Of course, this can be done by fiscal year instead of calendar year if desired.)

Objective 3. Use paper efficiently. Finding different ways to use paper and save paper will be achieved in the major paper objective attainment. Some of these different uses will be described later.

Objective 4. Recycle the used paper.
Recycling paper should be a given at most organizations today along with soda cans, water bottles, cardboard, batteries, computers/monitors/laptops/imaging equipment and paper, and ink cartridges.

If you do not recycle paper now, sell the organization on conducting a pilot. Get boxes and label them for paper (office paper, newspaper, magazines, catalogs, etc.) and place throughout the facility. Get the janitors to pick up weekly and carry to a closed container from a vendor who picks up paper in your facilities' yard or outside storage area. Have the vendor weigh the paper when they arrive at their yard and send your organization a copy of what it weighted and date picked up. Record this info and publish the recycling results at least quarterly. The savings will be amazing and show real support of the environment.

IMPLEMENTING THE MAJOR OBJECTIVE:
REDUCE OFFICE PAPER USE AND SAVE ENERGY

The 10 steps to achieve this objective are shown below:

Step 1. Appoint or identify a champion for reducing office paper. Select a manager that is interested in championing the objective.

Step 2. Have the champion make an assessment of inefficient

paper use at the organization, draft a policy for saving paper, and develop program to meet that need. The champion does a quick assessment of how much paper is purchased and used yearly, how many copiers are using duplex copies, and through interviews determines if files are kept in paper even though they exist electronically. Make an estimate of how much can be reduced in a 12-month period.

Step 3. Get top management approval for program and policy. Write an objective and organization policy and get management's approval. Communicate the policy throughout the organization.

Step 4. Organize a team to support program. Put together a cross-functional team and include some heavy paper users on the team. Prepare an agenda for each meeting and minutes of the meeting. Go over objective and the organization policy at first meeting. Establish roles of the team and expectations.

Step 5. Develop systems to track paper use and its impact. Develop a process to track paper purchased and used. This process will vary slightly at each organization. Some organizations have counters at their copiers or sheets to be annotated to provide a record. A problem could present itself in the purchasing, if more than one procurement person orders paper for the organization. If so, then a monthly report from each to a team member will be needed.

Step 6. Identify how reduction occurred or can occur, prioritize saving measures, and communicate to all participants. Brainstorm what measures can be employed to save paper. Prioritize the list and put into PowerPoint awareness training or place close to organization's printers and imaging equipment.

1. Print more pages per sheet (at least two pages a sheet and for handouts 4-6 pages per sheet).

2. Provide one copy for several readers.

3. Duplex copy = double-sided (your copiers should allow for this).

4. Put on bottom of emails, "Don't Print."

5. Use electronic files when possible.

6. Use a single-space format for reports.

7. Save work on CDs or flash drives.

8. Be selective in printing attachments.

9. Make fewer copies.

10. When copying, reduce size.

11. "Think Before You Print."

12. Use used paper for scratch paper or notes.

13. Use macros to shorten reports and fill out electronic forms.

Check with customers and ask if electronic copy would be okay instead of paper.

Step 7. Consider a pilot project. A pilot is good for a large organization with hundreds of white collar workers. It may also be a good idea for any size organization to test duplex printing.

In implementing duplex printing, note it is:

•   Mandatory in government;

•   All imaging machines (copiers, printers, and others) with capability should be set to duplex; and

•   If management does not allow, place a sign to duplex when you can and then reset the templates.

Designating one or two copiers and testing duplex printing and getting user's feedback as to the practicability and suitability plus how many sheets were saved works well. Use the feedback and savings information to decide on what to implement organizationally. A two-month pilot should suffice.

Step 8. Encourage involvement. Top management, paper champion, and paper reduction team should encourage all employees to participate. Explain the goal, the policy, progress and results to date. Show them what they can do to help contribute and keep everyone up-to-date with results. Use emails, posters, videos, and other communications media to sustain or increase the reduction efforts. Place reminders near the copy machine or at individual network printers or even at desktops.

- Use catchy slogans such as,
  "Do you really need to print that?"
  "Stop and think: do you really need extra copies?" or
  "Do you know how many sheets of paper you used last month?"
  "The average sheets used per employee are __?"
  "Can we reduce that average?"

- Track the personal printing footprint in your office.

- Develop systems that allow staff to measure how many print copies they personally used in a month. The amount will surprise everyone.

- Hopefully, this knowledge will motivate people to reduce their personal paper footprint.

Step 9. Track results over time and communicate to include success stories.

- Track paper used monthly, break down by department, and give a total for company or organization.

- Calculate percent paper reduced monthly.

- Show the environmental savings, the cost saved, and other facts.

- Track pounds of paper recycled monthly and their environmental impact.

Step 10. Continuously improve amount of sheets reduced, environmentally preferable paper, and reduce waste

Objective 1. Buy 30% recycled paper
- Actions:
  1. Put requirement in the specs.
  2. Randomly inspect to ensure the paper meets the requirement.
  3. Test your imaging equipment to how high a recycled content (30+) your equipment can tolerate.

Objective 2. Reduce amount purchased.
- Measure the pallets, boxes, and or reams of paper purchased.

Objective 3: Use paper efficiently. Use electronic files and duplex.

Objective 4: Measure the pounds of paper recycled.
*   Measurements show how effectively we are reducing use.
*   Pounds recycled verify that countermeasures are working.
*   Amount recycled will be less.
*   Using less paper.
*   Saving more trees.

Each month, count how many realms of paper were purchased and record the number sheets copied by imaging machines. The recycling company measures the paper that is recycled monthly, once they get to their scale at their recycling yard. Determine how many trees were saved, how many kWh were saved, how many gallons of water were saved, the reduction of air pollution, and the cubic yards of landfill saved. Calculate how much the organization saved by not having to purchase the additional paper. Communicate how well the program is going to motivate all participants and possible participants. Also, the cost savings keeps management supportive.

Convert paper saved into pounds saved. During this year, you used:

$$200 - 4 \text{ cases} = 196 \times 10 \text{ reams/case} = 1960 \text{ reams}$$
$$\times 500 \text{ sheets/ream} = 980,000 \text{ sheets in 12 months}$$
(an average of 3,920 sheets per person per year
or 81,666.67 sheets per month).

During last year, you used:

$$310 - 8 = 302 \text{ cases used.}$$
$$302 \text{ cases/year} \times 10 \text{ reams/case} \times 500 \text{ sheets/ream} =$$
$$1,510,000 \text{ sheets/year}$$
(sheets savings = 1,510,000 - 980,000 = 530,000)

Weight Saved = 530,000 sheets saved divided by 500 sheets/ream times 5 pounds per ream = 5,300 pounds of paper saved this year .

Next, convert to tons saved by 5,300 saved/2000 pounds/ton = 2.65 tons saved.

Answers:

How many trees were saved? 2.65 × 17 = 45 trees.

How many kWh were saved? 2.65 × 4,000 = 10,865 kWh saved.

How many gallons of water were saved? 2.65 × 7000 = 18,550 gallons saved.

The amount of air pollution was reduced by 2.65 × 60 = 159 pounds. As for landfill, 2.65 × 3.3 = 8.75 cubic yards were saved.

How much money was saved? 5,300 pounds divided by 5 pounds = 1,060 reams multiplied by $6.49 per ream = $6,879.40.

Was it Worth It? You Bet it was!

Now, set new targets and develop new action plans.
- Communicate to all, recognize excellent achievements.
- Keep champion and team in place and performing.
- Keep all management, supervisors, employees, and contractors informed of what has been successful, the results, and the savings.
- Accept new ideas and give credit to people's ideas where they deserve them. Recognize and reward achievements.

Chapter 12

Energy Centered Maintenance (ECM) & Energy Centered Projects (ECP) in Data Centers

ESTABLISH AN ENERGY REDUCTION TEAM

Form a cross-functional data center energy reduction team and elect a team leader. The team will be needed to reduce energy use in a data center. Team members should include the facility manager, assistant data center manager, energy manager, data center operating supervisor, a corporate engineering/facilities representative, and a corporate IT representative. There will be ECO objectives and targets, ECM activities and energy projects (ECP) for the team to develop and sell to management for funding.

KEY DEFINITIONS

Some key definitions that each team member should understand are:

CRAC—Computer room air conditioning.

PDU—Power distribution units—distributes power to servers, performs power filtering and load balancing, and provides remote monitoring and control.

Plenum—Concealed area that enables hot air flow back to CRAC.

PRO DC—A software program developed by DOE to identify energy savings in a data center.

PUE—Power use effectiveness = total facility power/IT equipment & operations use in kW rack-container that holds the servers.

Raised floor—enables the cool air or cold water cables to cool the servers.

Server—a computer that computes or provides service to a network or internet.

UPS—Uninterruptable power supply—provides backup power in case of an interruption to main power supply.

ENERGY TEAM ACTIVITIES

Next, the energy team should understand more about data centers and their characteristics. Some key data center characteristics are:

➢ Data centers are normally in operation 24/7/365.

➢ Data centers have HVAC, chillers, cooling towers, CRACs, Backup power (uninterruptable power supply-UPS), boilers, generators, lights, servers, cables, PDUs, and other equipment.

➢ Need troubleshooting and monitoring all the time.

➢ Declared by organization as a critical mission facility.

➢ Can be in a facility by themselves or in a facility containing other functions or organizations. Can be very large or located in one small room.

➢ Power to the facility remains flat through the year (no seasonal changes as seen in other facilities) and costs around or more than $50,000 a month depending on size and equipment inside.

➢ Data centers use around 2% of energy consumed in the U.S. yearly.

Servers are the essential mission equipment with a large cast of supporting equipment mentioned above such as CRACs, UPS and others. There approximately 19 different types of servers. In a data center, it is not unusual to have many of these in the network. The other equipment supports the servers who accomplish the data center mission.

Step 1. Review energy past use and establish a baseline
Step 2. Determine what the PUE is.
Step 3. Do the data center quick test
Step 4. Identify how energy can be saved
Step 5. Develop an action plan
Step 6. Perform payback analyses on projects
Step 7. Sell projects to management
Step 8. Track projects and monitor PUE and other metrics changes

IDENTIFYING ENERGY WASTE

"Let's overbuild just in case we need it" was for a long time what builders of data centers did at the owner's request. Doing that costs a lot of money as well as wastes energy.

Servers are not the only equipment in data centers that use and consume energy. Industrial cooling systems, circuitry to keep backup batteries charged and extensive wiring all consume energy. Hot air mixing with the cold air, thus requiring the cold air to be cooled again, is a significant energy waste.

In a typical data center, those losses combined with low server utilization can mean that the energy wasted can be as much as 30 times the amount of electricity used to carry out the basic purpose/mission of the data center.

Energy Team Steps
Step 1 is to establish a baseline.

☐ A baseline is established from historical data for an existing data center. It is good to have at least 3 years of data. For facilities less than 2 years, use energy modeling techniques. The baseline does not have to be last year, but any of the last 3 years. The year that appears to be the most typical should be the baseline year.

☐ The baseline enables comparison later when energy savings countermeasures have been implemented. The comparison shows how much energy savings has been achieved.

Figure 12-1 shows typical utility use or consumption for a data center. Notice it does not have the seasonal summer and winter peaks.

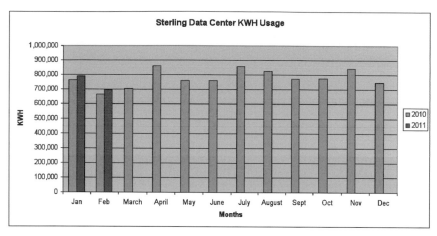

Figure 12-1. A typical data center Electricity kWh Usage

Step 2 determines the power usage effectiveness. This will give the energy team an indicator of how much the data center can be improved.

- Who publishes measures for data centers? Answer: ASHRAE, Energy Star, and Green Grid.

- What is the most popular metric or energy performance measure now? It is PUE. Power usage effectiveness = total facility power (kW)/IT equipment power (kW).

- A PUE of 1.6 to 1.3 is good, with 1 being the goal. The lower the number, the larger is the percentage of total power furnished to the facility that is used by the data center operations. (80% of the power furnished used by the IT equipment would be very good. This is the DCIE indicator whose reciprocal is the PUE. 1/.8 = 1.25)

Facebook's new data center in Forest City, NC, has a PUE between 1.06 to 1.08.

Estimating PUE When Data Are Not Available
Total facility kW comes from the meter where the electricity enters the building. It will be normalized by measuring the data center's square footage and dividing that by the building's square footage

where the kW entered and was measured.

The number of cabinets / racks will be counted. The IT equipment kW will be estimated by multiplying the number of cabinets times 4 kW per cabinet.

We can also check with 25 watts per square feet, or read the kW used by the UPS if they are totally dedicated to IT equipment and operations.

Step 3. Do the quick test. Walk in the area where the servers are located.

Quick Test: Is it cold? If yes, then too much energy is being used and can be saved or reduced.

Step 4. Determine how can we save energy? Start with cold aisle containment if you have not already done so. Arrange the servers in rows receiving cold air and exhausting in the next row.

If you have several racks of servers, use cold aisle/hot aisle concept. Separate cold and hot air.

Then analyze your equipment. Normally, data centers' cooling equipment has excessive capacity and is not operated at temperatures lower or higher than ASHRAE requirements due to the fear of losing servers and their contents.

Consolidating various applications onto less servers has a significant impact on energy efficiency by reducing both server energy consumption and cooling requirements. Normally, the result is energy savings of around 10 percent.

- Take unused services out of commission.

- Consolidate low-use servers to raise server utilization.

- Average temperatures vary and are very dependent on the quality of the hot and cold aisle isolation. These technologies are fairly new and the penetration is modest. Therefore most data centers are still cold. You would like to see the cold aisle temperature in the upper 70's and the hot aisle temperature over 100°F, however this depends on very good air management—*Dale Sartor, DOE*

- Don't mix hot and cold air. Many data centers mix hot and cold air, limiting the capacity and effectiveness of the cooling system. The countermeasure is placing air tiles in the cold aisle, and/or locating supply vents in the cold aisle and return vents in the hot aisle, along with a curtain or hard enclosure to contain the cold or hot aisle. Not only does this make the cooling more efficient, it raises return temperatures, allowing the CRAC units to operate more efficiently. Containing hot air removed from servers and cold air supplied to them increases efficiency and reduces power consumption.

- Before contracting the project, coordinate with local fire marshal, and keep him or her involved until project is complete.

- Variable frequency drives (VFDs) can improve the energy efficiency of cooling equipment by enabling the systems to run at a lesser percentage of the motor speed. This will result in less power used and will limit over-cooling in the facility.

- Use economizer, if feasible, to bring outside air in. Investigate each cooling component to ensure improved efficiencies.

Energy Star uninterruptible power supply (UPS) systems that deliver more than 95 percent efficiencies at partial loads should be normal. The single largest power user in a data center is the UPS. Energy Star-rated, and an on-line battery based UPS system where utilities provide high enough power quality, may lower the energy costs.

Step 5. Develop an action plan.
   Develop an action plan:
- What is going to be done?
- Who is going to do what?
- When it is going to be completed?
- The why and where are self evident.

Step 6. Perform payback analysis on each project.
   For each project, estimate the cost and the benefits in kWh reduction per year.
   Calculate the payback period.
   For the actions and projects with a payback period of 3 years and under, develop either a force field analysis or barriers and aides.

Step 7. Sell projects to management.
   Use barriers and aides or force field analysis.

Develop a presentation with all the actions/projects recommended and the kWh or Btus reduced by each. Put in this order:
➢ Actions costing no dollars or just a few.
➢ Projects with payback period of 3 years or less.
➢ Projects with payback period greater than 3 years.

Step 8. Track projects and metrics.
• Track PUE.
• Track cost vs. benefits achieved.
• Track kWh consumed monthly.
• Monitor installation of items per schedule.

Following the above steps and focusing on the possible options can save or eliminate energy waste up to 30% of total energy used by the data center.

Chapter 13

Building Your
Energy Reduction Plan

QUICK METHOD

The energy reduction deployment checklist contains the essentials to building an efficient and effective energy reduction deployment plan. Review the checklist with your energy team and others that you feel have a part in reducing energy at your organization. For each "no" you check, place the thrust of the question on an action plan as the "what." Then answer "who" will be the responsible person. Then "when" the task is expected to be accomplished. Any pertinent information should be included in the "remarks" column. The "status" column is to be used later when the energy champion, energy manager or energy team reviews the action plan for status. If the plan is on track, place "green" as the status. If it is just a little off track or behind, then place "yellow" in the status column. If it has not been started but should have been, or if a lot of effort is needed to get it on track to be completed when expected, and then place a "red" in the status column.

Figure 13-1. Energy Reduction Program Checklist

Yes___ No___ 1. Has top management established a corporate energy or electricity reduction goal?

Yes___ No___ 2. Has top management appointed a management representative, an assigned an energy manager, or both?

Yes___ No___ 3. Has the management represented established a cross-functional energy team?

Yes___ No___ 4. Has the energy team started meeting at least monthly and established their team roles and responsibilities?

Yes___ No___ 5. Has top management developed an energy policy and communicated it to all personnel?

221

Figure 13-1 (*Cont'd*). Energy Reduction Program Checklist

Yes___ No___ 6. Does the energy policy include the corporate goal, management's expectations in reducing and managing energy consumption, and their renewal energy goals?

Yes___ No___ 7. Has the energy champion and his appointed team completed the energy walkthroughs?

Yes___ No___ 8. Has the energy champion and the energy team established an energy or electricity baseline?

Yes___ No___ 9. Have the energy performance indicators been selected and a data collection plan developed?

Yes___ No___ 10. Has the energy champion and energy team developed an energy plan for at least a 3-year period?

Yes___ No___ 11. Has the energy team established at least one objective and target and action plan and started implementation?

Yes___ No___ 12. Has an energy awareness and conservation plan been developed and training provided to all personnel?

Yes___ No___ 13. Has a communications plan been developed, approved, and followed?

Yes___ No___ 14. Has an emergency energy plan been developed and approved to include sufficient backup for critical operations?

Yes___ No___ 15. Has a procurement policy been written and included in an organizational procedure?

Yes___ No___ 16. Has significant contribution including reducing energy waste from ECO, ECP, and ECM been accomplished?

Yes___ No___ 17. Have the low hanging fruit actions been accomplished or at least planned?

Yes___ No___ 18. Has an energy waste "identify and eliminate" mindset been achieved for all organizational personnel?

Yes___ No___ 19. Were human factors and thinking used in energy planning, development, and implementation efforts?

Yes___ No___ 20. Has ECM been implemented?

Yes___ No___ 21. Does the organization use a computerized maintenance management system to schedule, record, and manage the ECM items completion?

Yes___ No___ 22. Has an IT power management program been implemented at the organization for computer monitors?

Yes___ No___ 23. Has an IT power management program at the organization been implemented for CPUs?

Figure 13-1 (*Cont'd*). Energy Reduction Program Checklist

Yes___ No___ 24. Has a metering plan been developed to provide essential data consumption data?

Yes___ No___ 25. If the organization has a data center, has the energy waste been identified and have plans to eliminate or minimize energy waste been developed?

Yes___ No___ 26. Are the key documents and records included in a centralized document control system and kept current and accessible to the organization's personnel?

Yes___ No___ 27. Are the energy performance indicators kept current, posted or made available to all organization personnel at least quarterly?

Yes___ No___ 28. Is the energy team using critical success factors or the energy reduction checklist to accelerate energy reduction program efforts?

Yes___ No___ 29. Does management review the energy reduction program achievements, current objectives and targets, issues and barriers at least annually?

Yes___ No___ 30. Is there any evidence that some processes used in reducing energy consumption and cost have been improved since the program started?

Yes___ No___ 31. Is there any evidence of progress towards meeting the corporate goal?

Yes___ No___ 32. Are the employees and contractors kept informed of progress and results of the energy reduction program?

Yes___ No___ 33. Are the employees and contractors involved in energy conservation efforts?

For the first three questions of the checklist, assume the answer was no. Then the action plan would look like Figure 13-2, Action Plan.

THE PROCESS

Figure 13-3 shows this process in a simple flow chart.

Using this process, place the activities that were answered with "no" onto a Gantt chart and you have your implementation plan or roadmap.

What (Question 1-33)	Who	When	Status	Remarks
1. Goal	Top Management (CEO)	By Sept. 30, 2014		Strategic Sustainability Council
2. Management Representative	Top Management (CEO)	By Oct. 10, 2014		Becomes the Energy Champion
3. Energy Team	Mgt. Rep.	By Oct. 31, 2014		Cross-Functional team with one or more team members from Facilities or Engineering or both.

Figure 13-2. Action Plan

TRADITIONAL METHOD

Using this book as a guide, identify what needs to be accomplished at your organization to achieve energy reduction. Review each component (ECP&D, ECW, ECO, ECP & ECM) and select from each what needs to be done at your organization. Then use the action plan format if completed. Next summarize into an energy plan using format previously outlined. Remember, do the walkthroughs and the identification of objectives and targets. The generation of new items or things to do keeps the energy team motivated and productive.

Organizations' energy plans come in a lot of different formats. There is no single right way to format an energy plan. Its purpose is to formulate a plan that can be used to communicate to the organization the actions planned and the approved by management to accomplish the energy reduction goal. There are a few recommended areas or contents. They are:

- About the organization—mission & vision

- What the plan includes and why

- Energy review and profile

- The established energy reduction goal

- Scanning the environment

- Organization's energy policy

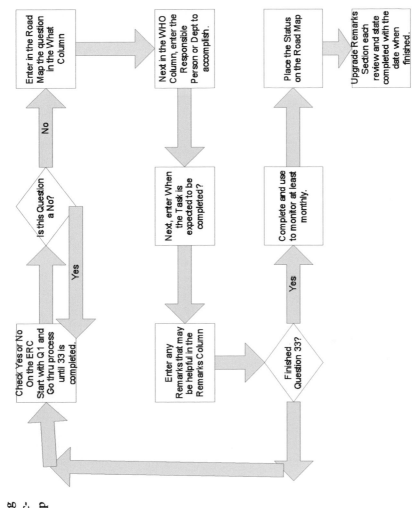

Figure 13-3. Building Your Energy Reduction Roadmap

Figure 13-4. Energy Reduction Deployment Implementation Plan

ID	Task Name	Start	Finish	Duration
1	Task 1 Est. Corporate Goal	6/13/2014	6/30/2014	12d
2	Task 2 Appoint Mgt. Rep.	6/13/2014	10/10/2014	86d
3	Task 3 Est. X-F Energy Team	10/10/2014	10/20/2014	7d
4	Task 4 Energy Team Starts Meetings	10/20/2014	10/24/2014	5d
5	Task 5 Team Establishes Roles & Responsibilities	10/24/2014	10/31/2014	6d
6	Task 6 Develop Energy Policy	10/31/2014	11/14/2014	11d
7	Task 7 Communicate Energy Policy to all personnel	11/14/2014	12/12/2014	21d
8	Task 8 Establish Walk Through Team	11/14/2014	12/19/2014	26d
9	Task 9 Do the Walk Throughs	12/19/2014	2/19/2015	45d
10	Task 10 Develop EPIs & Est. Baselines.	10/24/2014	11/24/2014	22d
11	Task 11 Develop Emergency and Contingency Plan	10/24/2014	1/23/2015	66d
12	Task 12 Est. Objectives & Targets & Action Plans	10/24/2014	11/24/2014	22d
13	Task 13 Develop Energy Awareness Training & Provide to all Personnel	10/24/2014	2/24/2015	88d
14	Task 14 Develop Energy Conservation Training & Provide to all Personnel	10/24/2014	3/24/2015	108d
15	Task 15 Develop Communications Plan	11/13/2014	1/15/2015	46d
16	Task 16 Develop Energy Plan	10/24/2014	1/23/2015	66d

Timeline columns: Q3 14 (Jul, Aug, Sep), Q4 14 (Oct, Nov, Dec), Q1 15 (Jan, Feb, Mar), Q2 15 (Apr, May, Jun)

#	Task	Start	End	Duration	
17	Task 17 Est. a Procurement Policy & Procedure & Communicate to all	10/24/2014	12/10/2014	34d	
18	Task 18 Determine Sufficient Contribution	10/24/2014	12/31/2014	49d	
19	Task 19 Pick the Low Hanging Fruit	12/31/2014	4/30/2015	87d	
20	Task 20 Implement ECP	2/19/2015	3/19/2015	21d	
21	Task 21 Implement ECM	2/19/2015	5/19/2015	64d	
22	Task 22 Place ECM onto CMMS & Schedule the activities	5/19/2015	5/19/2015	1d	
23	Task 23 Implement IT Power Management	6/15/2015	8/14/2015	45d	
24	Task 24 Implement Reduce Office Paper Reduction	7/1/2015	12/31/2015	132d	
25	Task 25 Id Energy Waste in Data Center & Develop Fix for them	8/13/2015	9/11/2015	22d	
26	Task 26 UseCSFs	10/24/2014	10/23/2015	261d	
27	Task 27 Conduct Management Review	9/15/2015	9/15/2015	1d	
28	Task 28 Establish central Documentation System	6/13/2014	12/12/2014	131d	
29	Task 29 Do FAME Visits	1/13/2015	7/13/2015	130d	
30	Task 30 Continuously Improve Operations, Activities, Processes, & Reduction of Energy	6/13/2014	1/6/2016	409d	

- Selecting the right strategies to achieve the energy policy and goal
- Roles and responsibilities
- The energy champion & how to contact
- The energy team, its members and how to contact
- Objectives
- Energy metrics and baselines
- Energy conservation
- Communications methods
- Funding sources/programming avenues
- Three- to five-year plan

Other possible topics are technology, renewable energy, market forces, opportunity assessments, strategies, sustainability, commissioning, re-commissioning, education and training and awareness, energy demand, water reduction, awards and incentives, verification and validation, and other areas. Make the energy plan yours. Let it communicate what each member needs to do to contribute and what measures need to be implemented, managed, and validated to reduce energy use, consumption and costs. The goal will not be achieved unless contributions are determined that when implemented achieve the desired amount. The next section will help to identify the most normal contributions. To reduce energy use further will require energy projects such as replacing or upgrading HVAC or retrofitting lights, new roof, etc.

DETERMINING CONTRIBUTIONS NEEDED TO ACHIEVE GOAL

There are many actions that can be implemented. The possible actions should be listed and estimates as to how much they could contributed toward achieving the goal should be developed. There is no scientific method to accomplish these estimates, but the procedure used should be able to be replicated later if needed. To shorten the number of estimates, the actions' priorities can be determined first, and only the high priority actions' contributions can be estimated.

The actions with 8 total score or more will receive estimates of contribution. They will be included in the energy plan to be accomplished during next 3 years.

Figure 13-5. Priorities of Actions

Actions	Feasibility 1-5	Ease of Implementation 1-5	Effective 1-5	Total Score	Contribution toward Goal
1. Implement an Energy Conservation Program	5	5	5	15	
2. Upgrade the Building Automation System and train the operator	2 (lack of funds)	2	4	8	
3. Implement IT Power Management	5	3	5	13	
4. Implement Energy Centered Maintenance	5	2	5	12	
5. Install Occupancy Sensors	5	4	3	12	
6. Seal the windows and doors	4	4	3	11	
7. Turn the thermostat to 78 degrees in summer and 68 in winter	5	3	5	13	
8. Change HVAC filters. Set AC Thermostats to Auto.	5	5	2	12	

(Continued)

Figure 13-5 (*Cont'd*). **Priorities of Actions**

Actions	Feasibility 1-5	Ease of Implementa-tion 1-5	Effective 1-5	Total Score	Contribution toward Goal
9. Set Hot Water Heaters to 120 degrees	5	5	2	12	
10. Rotate Fans for right season	5	5	1	11	
11. Replace windows with double paned ones	3	3	2	8	
12. Increase ventilation to areas of need	3	3	2	8	
13. Add insulation to attics and walls	3	3	2	8	
14. Install Big Fans in large warehouse areas	2	1	4	7	
15. Upgrade boilers with new sensors and switches	3	3	3	9	
16. Change T-12s lights to T-5s	3	3	4	10	
17. Replace roof with green or white roof.	1	1	5	7	

Figure 13-6. Estimated Contributions

**First, a WAG is done. It will save about 5% of the lights or HVAC. The mix of the total electric bill is HVAC 42%, lights 22%, office equipment 20%, and other 16%.

Action	Estimated Contribution	Remarks
Implement an Energy Conservation Program	2 %	**5 % Elect. .05X.22=.019 .05x.16=.008 Total=.019+.008=.0198 or 2%
Implement IT Power Management	0.7 %	Could be higher depending on number of computers and monitors
Implement Energy Centered Maintenance	10%	
Install Occupancy Sensors	.7 %	Will Cost a little money
Turn the thermostat to 78 degrees in summer and 68 in winter	2.1 %	Will require creating a positive culture
Seal the windows and doors	.84 %	
Change HVAC filters. Set AC Thermostats to Auto.	.4 %	
Set Hot Water Heaters to 120 degrees	.2%	
Rotate Fans for right season	.2%	
Upgrade boilers with new sensors and switches	1.5 %	
Change T-12s lights to T-5s	1.21%	Will require significant funding
Upgrade BAS & Train Operator	3.2 %	Will require significant funding
Total	23.05	

Implementation of energy conservation program will impact lighting (.22) and HVAC (.42), therefore multiply 5% or .05 times each and add for the total. Do the same for each action.

The energy team decided that they should not include the changing of the T-12s to T-5s nor the upgrading of the BAS since they both will require considerable funding. This makes 23.05 − 3.2 − 1.21 = 18.64%. The corporate electricity goal is 15%, and the identified contribution is 18.64 or 3.64% over. The energy team members felt this was a good cushion to have since something could go wrong with other actions.

SAVINGS OR COST AVOIDANCE VERIFICATION

If your electric bills showed that in 2012 (baseline year) your kWh consumption was 18,575,000, and for 2013 the kWh consumption was 17,895,344, did you save electricity? In the latter part of 2012, your organization implemented an energy conservation program, conducted energy awareness training and implemented both IT power management and a drive to reduce office paper projects. The energy team expected to see some savings. They were pleased to show the results to the energy champion. The energy champion asked if those were real savings. The team leader said, "Yes, our electricity cost in 2013 is less than that in 2012."

The energy champion replied that he had received a presentation the other day on SEP (superior energy performance) and their measurement and verification protocol analyzed the year-end totals for weather adjustment and production adjustment and maybe others. "First, let's see if the weather had any impact on our consuming less electricity in 2013. Try using www.freeenergymanagertool.com. They have a lot of free information, data and training. Let me know what effect, if any, the weather had on the difference in kWh consumption for the 2 years."

The energy team tried the recommended website and selected www.weatherdatedepot.com. First, the team felt that average temperature may be enlightening to the possible weather impact.

Table 13-1. Average Monthly Temperatures

Year	Jan	Feb	Mar	Apr	May	Jun	Jul
2010	42	40	53	64	73	85	84
2011	40	47	59	68	71	85	90
2012	47	50	62	68	75	82	85
2013	46	49	53	60	70	80	82
2014	42	43	52	63	72	81	81

Aug	Sep	Oct	Nov	Dec	Total	Average
87	77	65	56	46	772	64.33
91	77	65	55	44	792	66
84	77	64	56	48	798	66.5
86	78	65	50	39	758	63.17
					434	62

The average temperature for 2013 was 63.17°F and was lower than 2012 at 66.5°F. Based on this alone, weather could have been favorable and help cause the usage avoidance or savings.

The terms cooling degree days and heating degree days seem to be useful in verification and measurement. Cooling degree days (CDD) is the number of degrees that the average temperature is above 65°F and people start using their air conditioners to cool their facilities. The heating degree day is the number of degrees that the average temperature is below 65°F and people start using heat in their facility.

Table 13-2. Monthly Degree Day Comparisons (Station: DTOX)

Month	Base Year (2012)			Comparison Year (2013)			Comparison Percentages		
	HDD	CDD	TDD	HDD	CDD	TDD	HDD	CDD	TDD
January	403	8	411	441	18	459	9%		11%
February	308	19	327	315	11	326	2%		0%
March	79	147	226	234	45	279	196%	-69%	23%
April	1	250	251	108	132	240		-47%	-4%
May	0	489	489	29	361	390		-26%	-20%
June	0	668	668	0	618	618		-7%	-7%
July	0	805	805	0	703	703		-12%	-12%
August	0	754	754	0	808	808		7%	7%
September	0	529	529	32	577	609		9%	15%
October	88	226	314	35	192	227	-60%	-15%	-27%
November	169	72	241	315	26	341	86%		41%
December	411	39	450	625	3	628	52%		39%
Annual Total	1459	4006	5465	2134	3494	5628	46%	-13%	3%

The minus sign shows 2013 was worse, and the plus sign shows 2012 was better. The cumulative cooling degree days are shown in Figure 13-7.

In 2012, there were 4,006 total CDDs compared to 3,494 in 2012. The fewer CDD helped reduce electricity consumption at our organization in 2013, but how much? Did we achieve any real savings in 2013 or not? We thought we had saved 679,646 kWh in 2013 compared to 2012 with a cost savings of 679,656 kWh × $.07/kWh = $47,575.92. The energy team was baffled. How do we determine how much the weather helped?

Weather normalization techniques are usually based around using regression analysis of past energy consumption data, a technique that is frequently used with degree days to:

Figure 13-7. Cumulative Cooling Degree Days (Station: DTOX)

Year	Jan	Feb	Mar	Apr	May	Jun	Jul	Aug	Sep	Oct	Nov	Dec
2010	1	1	21	179	609	1363	2127	2990	3527	3705	3774	3788
2011	0	64	186	457	822	1587	2531	3508	4021	4225	4297	4300
2012	8	27	174	424	913	1581	2386	3140	3669	3895	3967	4006
2013	18	29	74	206	567	1185	1888	2696	3273	3465	3491	3494
2014	2	6	52	224	612	1248	1906					

- Attempt to identify signs of waste in past energy-consumption data.

- Assess recent energy performance by comparing recent consumption with a past-performance-based estimate of expected or forecasted consumption. In particular, this process is often used to quantify the savings from improvements in energy efficiency.

Also, there have been several building energy simulation models to analyze weather's impact on energy consumption. Most of these techniques have not been very accurate. The simple methods have done as well as the sophisticated ones. The energy team felt a simple method would suffice now that the CDD and HDD data are available. They first checked the year's average temperatures:

2012   66.5 degrees F
2013   63.17 degrees F
       3.33 degrees F $3.33/66.5 = .05 = 5\%$ decrease

Next, the team looked at the HDD and CDD.

Table 13-3. Yearly CDD & HDD

Year	HDD	CDD	TDD
2012	1459	4006	5465
2013	2134	3494	5628

The number of days in 2013 requiring heat significantly increased, while the days to cool significantly decreased. The total degree days in 2013 increased 3%. The air conditioning and heat comprised 42% of the total energy used by the organization: lights 22%, office equipment 20%, and other 16% comprised the rest. If 2012 was less than the percentage multiplied by 42%, the result multiplied by the total reduction in kWh consumption (savings) from 2012 to 2013 would give an appropriation of the actual kWh saved. However, with the TDD in 2013 being higher than that of 2012, the 67,656 kWh saved was actually realized and probably some additional that will not be estimated and counted.

The energy team presented this to the energy champion who accepted the findings, but recommended, for next year, to use the regression analysis techniques like DOE has developed for their SEP measurement and verification protocol.

CREATING THE ORGANIZATIONAL CULTURE— AN ASSESSMENT

An organizational culture dedicated to helping reduce energy consumption and cost is much desired to reduce energy and to continuously improve the processes and deployment. An organizational culture assessment is shown below.

Creating an Energy Efficient Culture, GLUE (Great Leadership in Using Energy)

➢ Leadership provides the GLUE that keeps the organization focused and productive.

➢ Without top management commitment and involvement, culture does not become one of finding energy waste and eliminating it.

Two questions of importance: What is it? Can we favorably impact it?

➢ **Top Management/Leadership**
  1. ___ Considered to be like parents (promotes honesty and openness).

  2. ___ Considered to be pacesetters (hard drivers and completely in charge).

3. ___ Once our leaders decide on a course of action, it will get done.

4. ___ They know what they want when we show them.

> **Strategic Planning (Vision, Policy, & Goals) and Emphasis**
1. ___Does your organization have a strategic plan with a vision and corporate goals?

2. ___ Is there a corporate goal to reduce energy cost and usage?

3. ___ Is there an energy policy? If no, is there an environmental policy, quality policy or safety policy? Does the vision include reducing energy or operating cost or productivity?

4. ___ Do you have a strategic council, leadership council or something similar in place?

> **ISO Standards**
1. ___ Have you implemented or are you thinking about implementing ISO 50001 Energy Management System (EnMS)?

2. ___ Have you implemented or are you thinking about implementing ISO 9001 Quality Management System (QMS)?

3. ___ Have you implemented or are you thinking about implementing ISO 14001 Environmental Management System (EMS)?

4. ___ Have you implemented or are you thinking about implementing OHSMS 18000 Safety Management System (SMS)?

5. ___ Has the organization appointed any management representative to lead an ISO standard implementation effort and has she or he defined the roles and responsibilities of those involved?

> **Organization Characteristics**
1. ___ People in the organization are close and committed to the organization.

2. ___ Our organization is innovative and on the cutting edge.

3. ___ Our organization's primary emphasis is on production

and objectives attainment.

4. ___ Our organization's heart is policies, procedures, and work rules.

➤ **Values**
1. __ Our corporate values and principles are not known or not published.

2. ___ Our corporate values are published but not communicated well to all employees.

3. ___ Our corporate values are communicated to all our people, and we believe in them.

4. ___ Our corporate values and guiding principles do not make sense.

➤ **Communications**
1. ___ Our organization communicates very well and uses more than one media.

2. ___ Communications is not our organizations strong point.

3. ___ All important performance indicators including electricity kWh usage are kept current and posted where all can see.

4. ___ Management holds regular staff meetings, and energy usage is sometimes discussed.

➤ **Climate and/or Morale**
1. ___ The attitudes, feelings, and perceptions of individuals here are very positive.

2. ___ The morale seems to change whenever a major event or happening occurs such as a change in leadership.

3. ___ Most everyone is proud to be a part of our company.

4. ___ The company sucks. I have a job, and that is about it.

➤ **Training/Conservation**
1. ___ Our company focus on keeping their employees competent and up-to-date.

2. ___ We only have orientation training and occasionally some training the boss feels important.

3. ___ We have had some six sigma and lean training.

4. ___ We have had energy conservation training but do not practice it.

5. ___ We have had energy conservation training and actively participate in reducing costs.

6. ___ Power management has been implemented and we purchase only energy-saving computers and equipment.

➤ **Roles and Responsibilities**
1. ___ Has the energy team identified their roles and responsibilities?

2. ___ Does the staff know their energy reduction deployment roles and responsibilities?

3. ___ Has a management representative been appointed for energy management and reduction?

4. ___ Does your organization have a corporate energy manager?

5. ___ Is there an energy manager or someone responsible for energy at each of your organization's facilities?

➤ **Business Processes**
1. ___ Have the business core processes been identified and communicated at your organization?

2. ___ Do key processes have process owners?

3. ___ Have any processes related to energy reduction or management been mapped and communicated?

4. ___ Has ISO 9001 Quality Management System Been Implemented at your organization?
   • 8 Areas

   • Example: 6 energy team members took the survey

   • For each area, check your score and add for a total score.

How is your culture? In what areas do you need improvement (if any)?

- Top management Q1-5 Pts, Q2-5 Pts, Q3-7 Pts, Q4-0 Pts (17 points possible).

- Strategic planning Q1-7 Pt, Q2-7 Pts, Q3-7 Pts, Q4-6 Pts (27 points possible).

- ISO-Q 1-7, Q2-4, Q3-5, QA4-4, Q5-7 (27 points possible).

- Organizational characteristics Q1-7, Q2-7, Q3-5, Q4-3 (22 points possible).

- Values Q1-0, Q2-2, Q3-7, Q4-1 (10 points possible).

- Communications Q1-7, Q2-0, Q3-7, Q4-6 (20 points possible).

- Climate & morale Q1-7, Q2-5, Q3-7, Q4-0 (19 points possible).

- Training Q1-7, Q2-5, Q-3-5, Q4-2, Q5-7, Q6-7 (33 points possible).

- Roles and responsibilities Q1-4, Q2-5, Q3-7, Q4-6 (22 points possible).

- Business processes Q1-5, Q2-6, Q-3-7, Q4-7, Q5-7 (32 points possible).

Chapter 14

Drivers of Energy Reductions and Continuous Improvement & Verifying Results

WHAT IS A DRIVER?

A driver facilitates or enables a program to implement the plan quicker and more efficiently than if the driver were not present. A champion and his or her actions, leadership, a supportive activity, a policy or vision, an objective and target, a review, supportive actions such as speeches and providing resources, and other such actions that tend to motivate people and provide momentum to improvement initiatives are possible drivers. New major programs or systems implementation need drivers to maintain their momentum, management support, and employees involvement.

THE DRIVERS OF ECMS

First, leadership, their support and involvement, is very important to every improvement initiative. Appointing an energy champion to lead the energy reduction effort is an excellent move by management. Then establishing a cross functional team to address energy issues and opportunities for improvement is needed. The champion will head the energy team or select an exceptional team leader who will also report to him or her.

Developing objectives and targets (O&T) and action plans and designating an energy team member to be the responsible person for the O&T is a driver of significant importance.

Management, with the help of the energy champion, should develop a compelling energy policy and include implementing ECMS as a part of it will certainly help drive the program. Include identifying energy waste and eliminating or minimizing it as part of the organization's future plans. Making the energy reduction and implementing ECMS as a corporate objective will stimulate positive actions.

Energy indicators with the target or goal shown is a major driver. For everyone to know where we are as an organization and where we need to go will foster cooperation, involvement, and support. Keep the energy indicators current and make them visible to everyone.

Reviews (checks) of what the progress is, identifying and overcoming barriers to problems, determining if resources are adequate or if additional may be needed are some of the important functions of reviews. In ECM, reviewing the CMMS reports of accomplishment, backlog, and using the energy indicators to gauge results are excellent drivers of improvement. Project management including project reviews by the responsible person for the O&T, energy team reviews, and management reviews help identify where we are and what should be done or corrected; then set forth those actions. Self inspections done by the energy team and/or internal audits done by organization's auditors can help accomplish this also. Reviews are good. They are definite drivers of actions.

Of course, the huge cost of energy and the realization that whatever cost is saved can be spent on other important requirements is a driver. Using ISO 50001 Energy Management System either as a guide or to implement at the organization's facility or facilities is also a powerful driver. The drivers help progress increase and accelerate results achieved.

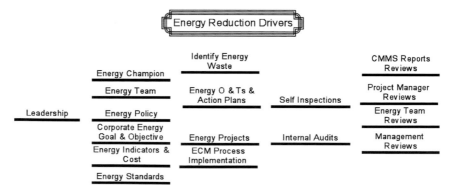

Figure 14-1. The Energy Drivers

Energy reduction and energy management are not normally onetime things; they are a continuous journey of improvements. The processes are analyzed and improved. The gains achieved in the energy reduction efforts must be maintained or some will be lost, because the maintenance activities were no longer occurring or new equipment has been introduced and ECM may not be implemented for that. Follow the wheel below for continuous improvement. Update the energy policy as appropriate and communicate to all. Develop new objects and targets and action plans and always have a few to keep the energy team active and productive. Implement them and do the reviews and corrective actions as needed. Accomplish Management reviews at least annually. Use the energy standard's inputs and outputs to ensure an efficient and productive meeting that covers all necessary materials and leaves out no important issues. Reward and recognize team, organizational sections, and individuals' contributions as appropriate.

Figure 14-2. Plan-Do-Check-Act Cycle

A BARRIER OR NON-DRIVER WHICH MUST BE CONSIDERED

What can occur that may impact determining if an action actually reduced kWh consumption or not? What can we do? An energy team justified changing the lights in the administrative areas from T-12s to T-5s. Top management funded the project when the team told them it would reduce the electric bill by 2.5% a year or by 7.5% in three years. The payback period was under 3 years. The project was installed by August 1, 2013.

The kWh for August 2013 was 280,800. The trend of 2012 and 2013 is shown in Figure 14-3.

Figure 14-3. The 2012 and 2013 kWh Consumption

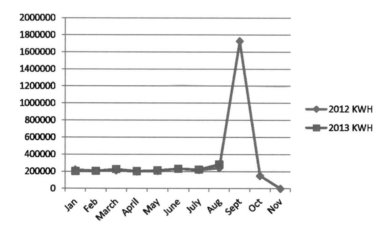

The energy team could not wait to see the results in September 2014, after the August 2014 electricity bill arrived with the kWh consumption. They had tracked the other months for 2014 and compared the results with 2013. What do you think the results showed? The total kWh consumption for 2013 and 2014 Jan-Aug was:

2013—784,927 kWh   2014—1,850,547 kWh

2014-2013 = 1850547 − 1784927 = 65,620 or 3.68% increase

The target was a 2.5% decrease for a year. Although this is only an 8-month comparison, it does not look like the goal is reachable. What happened? What do we tell management? Is this the end of our energy reduction program?

The energy team was concerned and some were very scared that the energy reduction effort was over. Management would be disappointed and may cancel further energy reduction efforts.

Discussion led to asking, "Did the new lights actually reduce energy use but something else caused the electricity usage increase?" "If so, what?" "What can we tell management that will be clear and hopefully acceptable?

"Did the new lights actually reduce energy use but something else caused the electricity usage increase?" The answer is yes. The energy team remembered the electric light vendor measured the volts and amps being used by both the T-12 lights and then the T-5s lights and found the T-5s lights used 10% less power. "If so, what?" The energy team and facilities/engineering checked projects and departments to see what was added between August 2013 and August 2014. They were surprised. The additions were:

1.   A 10% increase in servers occurred in Sept.-Oct 2013 in the data center.

2.   Two large vacant rooms used for storage were converted to office space including work stations with computers and monitors and over 60 people. An additional HVAC system was added on the roof to accommodate the needs for these two rooms.

3.   Three new conference rooms were added, and windows did not have film.

4.   Several motors were found running 24 hours a day when they only needed to run 4 to 7 hours a day.

5.   The chiller was not maintained properly and was using excessive electricity.

The energy team concluded that

1.   Presently, our electricity is read at one location. We cannot determine how much electricity was used for lighting, how much for the data center, for air conditioning, for heating, or for ventilation. We only know what was used for the building. Recommend we develop a metering plan to add additional meters or sub-meters and maybe have them come together in a console (electricity,

natural gas, and water usage) and determine its cost effectiveness prior to presenting to management for approval and funding.

2.   Any new addition that uses electricity or natural gas, we document when it was added and how much kWh or CF is estimated that it will use in a year.

Management appreciated the explanation given by the energy team and recommended they pursue their suggestions/recommendations.

This is a real barrier and must be addressed in the verification and validation of both small reductions and large ones from UESCs or ESPCs or any organization's energy projects or initiatives.

Chapter 15

ECMS Glossary

DEFINITIONS

Action Plan—The steps that must be taken to accomplish an objective and target or project showing who is going to do what and when. A Gantt chart is the most common form of an action plan.

Benchmarking—A method that enables one organization to compare their process with another's whose performance is much better, and to use the information to improve their process performance.

Best Available Techniques—Best means most effective. Available means can be used in your situation at a reasonable cost. Techniques includes what technology is used and the way in which installation or implementation is designed, installed, maintained, operated, and later decommissioned.

Best Practices—An accepted way done by a world class organization that is proven to be the best way of accomplishing a goal.

Brainstorming—An idea-gathering technique that uses group interaction to generate as many items as possible in a designated time. Quantity of ideas is the focus.

Charter—A written commitment by management detailing an improvement group's or team's authority, resource needs, including time and funds, to complete a purpose or assignment.

Check List—A tool that outlines key events to make something happen.

Check Sheet—A quality tool designed to collect data.

CMMS—Computerized maintenance management system.

Contribution—The estimated amount that a project or action will reduce the electricity or natural gas or any other energy use is called contribution.

Cost-benefit Analysis—Analysis to show both the costs and resulting benefits of a plan.

Cross-functional—A term used to describe people from several different functional organizational units or functions brought together on a team for a specific purpose.

Countermeasure—A potential solution that should eliminate or minimize a root cause to a problem.

Critical Success Factors—What you are paid to accomplish. Process or program will not succeed unless these factors are done properly.

Cultural Change—A major shift in attitudes norms, sentiments, beliefs, values, operating principles, and behavior of an organization.

Customer—Anyone for whom an organization provides goods or services. Can be internal or external to the organization. The next user of a product.

Defect—A non-conforming attribute.

Duplex Printing—Two-sided printing; i.e., print on both sides of a copier paper sheet.

Effect—An observable action, result, or evidence of a problem. It is what happens.

Energy Conservation Measures (ECMs)—A government form that shows projects that if funded would reduce energy and/or water use and provides a description of project and payback period or rate of return on investment.

ECM—Energy Centered Management—An energy management system that includes a "find energy waste" segment (energy centered waste) and three other segments: ECM (energy centered maintenance) a maintenance approach to keep equipment from using excess energy, ECO (energy centered objectives) a segment that fixes the waste by elimination or minimizing by energy team establishing objectives and targets with action plans, and a fix-it segment that includes energy projects that are contracted, including commissioning a facility.

ECM—Energy Centered Maintenance—A maintenance approach to prevent equipment using excessive energy.

ECO—Energy centered objectives where energy teams establish objectives and targets with action plans to address energy waste.

ECP—Energy centered projects where energy projects are contracted that will reduce energy use and consumption that includes commissioning of a facility.

ECW—Energy centered waste is a segment to find energy waste through walkthrough (observation), energy audits, and research, so it can be eliminated or minimized.

Energy Audit—An audit of the facility including infrastructure, building envelope, HVAC, other equipment, motors, exit signs, lightning, etc. to identify areas where energy use can be reduced. Result is a list of projects with the ROI for each.

Energy Champion—The management representative required by ISO 50001 EnMS.

Energy Efficiency—To accomplish products and services with less energy consumed.

Energy Intensity—The energy consumed in kBtus divided by the total gross square footage of the facility. This measure enables comparison with similar facilities.

Energy Model—Placing all energy use on a data sheet, analyzing it to determine where reductions may be possible.

Energy Variables—Variables that are correlated with energy use or can be used to normalize energy use for comparing with other organizations.

EPEAT—Energy product environmental assessment tool. A certification by a non-government organization that a product is energy friendly and meets their specifications.

Facilitation—A facilitator helps a team through a process or method, and makes it easier for them to achieve their purpose.

Fat Rabbit—The area where the most improvement can be achieved if action is taken.

Fishbone Diagram—A cause-and-effect diagram invented by Ishikowa to show causes and their effects. Used to identify root causes.

Five Whys—A technique for discovering root causes of a problem by repeatedly asking and answering the question why?

Flow Chart—A chart that shows inputs—the process (sequential work activities)—outputs and outcomes.

FMEA—Failure modes effects analysis, a tool to determine risk of a component or equipment to fail.

FMECA—Failure modes effect criticality analysis—The FMEA with an added criticality analysis.

Force Field Analysis—A graphical representation of barriers and aids to help sell an implementation plan to solve a problem.

Goal—A broad statement describing a future condition, or achievement without being specific about how much and when. The establishment of a goal implies that a sustained effort and energy will be applied over a time period.

Graphs—A visual display of quantitative data over time. Either a line chart, column chart, run chart, or bar chart.

Inputs—Materials, equipment, training, people, dollars, energy, facilities, systems, etc. needed to start a process.

ISO—International Standards Organization located in Switzerland that develops with nations standards for various functions, processes, and systems.

Kaizen—Continuous improvement in Japanese.

Kaizen Event—Usually a one-day event, with a group to improve a process.

Life Cycle Costing—The total cost incurred with a purchase and operation of a product until it is salvaged.

Mean—An average. A measure of central tendency of data (sum of total divided by number of observations).

Median—The middle point of data. A measure of central data.

Metrics—Indicators or measures overtime of a process whose objective is for improvement. Establishes a standard for management action.

Mode—The most recurring number in data.

Monitoring—Reviewing key indicators or equipment performance at specified times.

Multi-voting—A technique that allows a group to prioritize a large list down to three to five of the most important items.

Normalize—A method of making energy data or any data comparable with other similar facilities.

Objective—A specific statement of a desired short-term condition or achievement should include measureable end results to be accomplished within specific time limits. The how and when for achieving a specific goal.

Outcomes—Results from outputs. For example, customer is satisfied; savings resulted, cost avoidance obtained, etc.

Outputs—Products, material, services, information provided to customers (internal or external) resulting from a process(es).

Overall Equipment Effectiveness (OEE)—A method of breaking a plant, equipment, or process into availability, performance, and quality to determine their efficiency and effectiveness.

Pareto Chart—ABC analysis. A quality tool that identifies the most important or significant problem or opportunity. The 80-20 rule: 80% of the potential savings come from only 20% of the problems.

PDCA—The Deming or Shewart Cycle or wheel for continuous improvement.
Four stages: P—Plan: Plan what you are going to do.
D—Do: Implement Plan.
C—Check: Check the results.
A—Act: Take whatever action is necessary to get back on track to meet the target.

Plan—A goal or objective, metric(s), target(s) and a method and activities to accomplish.

Policy—A goal, objective, metric(s), and target(s).

Problem Solving Process—A structured process, when followed, will lead to solutions to a problem or processes. DMAIC is an example.

Process—A sequence of activities that produce a product, a service or information.

Process Control—Process is under statistical control.

Process Control System—System that identifies who does what (flow chart) to achieve an objective of a process. Also, identifies process metrics and outcome metrics.

Process Owner—Individual with responsibility of managing and controlling the process.

Product—A tangible output.

Power Factor—An adjustment fee caused by harmonics in the system inside a facility that forces the utility to have available more power than is used. The factor goes from 0 to 1 and anything under .94 incurs an adjustment fee on the facility's electric bill in some states.

Power Management—Placing computers, monitors, and laptops in a sleep mode when they have not been used for a specified time to save energy.

Preventative Maintenance—Routine maintenance accomplished in a specific time period to prevent equipment deterioration or failure.

Process Map or Flow Chart—A method using a diagram that shows input and the sequence of activities to turn the inputs into outputs such as a product, service or information.

Quality—Conformance to valid requirements.

Quality Assurance—A process of obtaining quality by measurement and analysis of work methods. Builds quality into the design phase—not inspect out.

Quality Improvement—A systematic method for improving processes to better meet client's needs and expectations.

Quality System—A chart that depicts for each business step of the company, what activities or processes with necessary information/data must be done to assure quality and business step effectiveness.

Quality Tool—An instrument or technique that supports the activities of process quality management and improvement.

Range—The highest point of data minus the lowest.

Root Cause—Why a requirement in a process is not being achieved.

Regression Analysis—A technique that allows for determining one variable's relationship with another variable or variables.

Reliability Centered Maintenance—A method to identify the critical parts of a plant or facility and ensure that they are always operable.

Root Cause Analysis—Analysis using a technique such as a fishbone diagram to identify the root cause(s) of a problem.

Self Inspection—An audit by a team using a checklist of the key elements to determine if they are in compliance and conformance to an ISO standard.

Server—a computer that accommodates one or more systems and is housed in a data center.

Six Sigma—A statistically based improvement methodology that strives to limit defects to 3.4 failures per million possibilities or six standard deviations from the mean.

Stakeholders—Individuals who have a stake in their organization's mission, change, and results. Customers, employees, suppliers, management, etc.

Standards—Factors usually established through statistics or measurement that serve as a basis for comparison.

Strategy—A determination of how you impact an objective to get where you want to be.

Strategic Direction—A vision of where an organization wants to go in the future or what it desires to be.

Strategic Objective—A corporate objective that supports the vision and strategic direction the company or organization has established.

System Boundary—A statement of what elements are included under the analysis being accomplished.

Target—A desired state or standard tracked through an indicator or the method or strategy that an objective is going to be achieved.

Task—Specific, definable activities to perform an assigned piece of work often finished in a certain time.

Team—A collection of people who are brought together to work on an objective or problem.

Team Leader—A person who leads a team through a process to achieve an objective or solve a problem.

Team Member—Serves on a team.

Teamwork—Works harmoniously towards achieving an objective or solving a problem.

Value Added—What happens in a process to change the inputs into outputs.

Vision—Senior management through analysis and customer voice determines where they want the company or organization to be in the long term (5 to 10 years).

Walk the Talk—Leading by setting a good example.

Chapter 16

Bibliography

BOOKS / ARTICLES

1. American Society of Heating, Refrigerating, and Air-conditioning, Engineers 1980. ASHRAE Handbook 1980 Systems.

2. Camp, Robert C. "Benchmarking: Finding And Implementing Best Practices That Lead To Superior Performance," Milwaukee, WI: ASQC Quality Press, 1989

3. Federal Energy Management Program: Electricity Reduction: www1.ere.energy.gov

4. Haasl, T., and K. Heinemeier. 2006. "California Commissioning Guide: New Buildings" and "California Commissioning Guide: Existing Buildings." California Commissioning Collaborative.

5. Howell, Marvin T., "Actionable Performance Measurement, A Key To Success," Milwaukee, WI, ASQ Quality Press, 2005

6. Howell, Marvin T., "Critical Success Factors Simplified, Implementing the Powerful Drivers of Dramatic Business Improvement," New York City, N.Y., CRC Press, Taylor and Francis Group, 2010

7. Howell, Marvin T., "Effective Implementation of ISO 50001 Energy Management System (EnMS), Milwaukee, WI, ASQ Quality Press, 2014

8. Howell, Marvin T., "Facilitating For Results—Expert, Manager, Mentor," New York City, N.Y., CRC Press, Taylor and Francis Group, 2014

9. ISO 50001 Energy Management System Standard, 2011, Requirements with Guidance for Use

10. ISO Energy Management Systems, Guidance for the Implementation, maintenance, and improvement of an EnMS, October 28, 2011, ISO/WD.2, Secretariat: ANSI

11. ISO 14001 Environmental Management System Standards, 2009

12. Mears, Peter, Ph.D., "Quality Improvement Tools and Techniques," McGraw Hill, New York, 1995.

13. ASME. 2011. ASME website. http://www.asme.org/search.aspx?searchText=EA&#page=1,category=STANDARDNew York: ASME. [EIA] Energy Information Administration. 2007.

14. Kowley, N., and A. Chittum. 2011. "Industrial Energy Efficiency Programs and Supporting Policies: A White Paper." Denver, Colo.: Western Governors' Association. McKane, A. 2011.

15. Scheihing, P., S. Schultz, J. Almaguer, et al. 2009. "Superior Energy Performance: A Roadmap for Achieving Continual Energy Performance Improvement." In Proceedings of 2009

16. ACEEE Summer Study on Energy Efficiency in Industry. Washington, D.C.: American Council for an Energy-Efficient Economy. [U.S. CEEM] U.S. Council for Energy-Efficient Manufacturing. 2011.

17. Superior Energy Performance program website. http://www.superiorenergyperformance.net. Washington, DC.

WEBSITES

1. aee online training.com
2. www.ase.org/efficiencynews
3. www.ask.com/ISO+Standards
4. www.ansi.org

5. www.doe eguide for EnMS.gov
6. www.eere.energy.gov
7. www.electricityforum.com/saving electricity html
8. www.google.com
9. www.iea.org
10. www.iso.org
11. http://www.iso.org/energy_management_system_standard
12. http://www.sei.ie/energymap/

Index

A

action plan 25, 35, 50, 70, 85, 88,
150, 151, 160, 163, 170, 218,
221, 223, 224, 241, 243, 247
audit 33
audit team 150
awareness 228
awareness training 172

B

baseline 54, 59, 67
brainstorming sessions 78

C

certification xv
computerized maintenance management system (CMMS)
247
reports 242
communication plan 27
communications 26, 27, 33
conservation training 26
continuous improvement 149
contribution 163, 170, 247
contributions 160
corporate electricity goal 231
corporate goal 141, 160, 236
corrective actions reports (CARs)
33, 37, 38, 150, 176, 180, 185
countermeasures 14
critical success factors (CSFs) 24,
25, 28, 29, 30, 223
cross functional 241
energy 139

energy team 3, 14, 51, 153, 157,
221

D

data collection plan 23, 154, 173,
174
Department of Energy's Superior
Energy Performance 1
deployment xi
documentation 33

E

efficiency 24
electricity intensity 49
energy action plans 145, 164
energy audit 14, 45, 51, 65, 96
energy-aware culture 157
energy awareness 14, 26, 65, 77, 91,
154, 159
and conservation program 101
conservation training 6
culture 164, 189
training 12, 35, 63, 76, 78, 140,
153, 193, 232
energy centered maintenance 14, 66
energy centered management system (ECMS) 72, 99, 139, 143,
150, 151, 179
energy champion 3, 14, 26, 33, 34,
41, 50, 51, 52, 54, 56, 69, 139,
141, 153, 157, 172, 173, 175,
228, 241, 249
energy conservation 27, 74, 91, 102,
142, 152, 154, 228, 238

efforts 223
measures 66, 76
plan 51, 154, 163
program 5, 27, 63, 67, 71, 85,
    140, 193, 231, 232
training 12, 35, 76, 172
energy efficiency (EE) xiii, 23, 24,
    48, 50, 66, 86, 200, 249
energy functional teams 72, 140
energy goal 168
energy indicators 242
energy management program 72,
    160, 164
energy management system xi,
    149, 157
energy manager 14, 193, 213, 221,
    238
energy metrics xiv
    and baselines 228
energy performance 223
    indicators (EnPIs) 14, 33, 54, 59,
    69, 72, 77, 139, 145, 154, 173,
    174
energy performance measures 57
energy plan 14, 33, 34, 41, 50, 60,
    61, 63, 69, 140, 153, 154, 164,
    167, 179, 224, 228, 229
energy policy 1, 26, 33, 34, 47, 48,
    50, 69, 139, 141, 153, 154, 164,
    167, 168, 221, 224, 236, 242,
    243
energy procurement plan 38
energy procurement policy 67, 196
energy profile 23, 67
energy projects 48
energy reduction xi, 76
    checklist 30, 223
    deployment 141
    deployment checklist 221

deployment plan 221
deployment process 33
efforts 50
goals 69, 119
initiative 78
program 178, 223
energy services performance con-
    tracts 12
energy team 3, 25, 33, 35, 41, 50, 52,
    54, 56, 60, 63, 64, 86, 91, 92,
    95, 141, 143, 150, 152, 165,
    170, 172, 173, 175, 176, 178,
    179, 193, 221, 223, 228, 231,
    232, 238, 241, 243
    cross-functional 91
    members 33, 69
    training 60
energy waste xi, 12, 14, 61, 62, 63,
    65, 95, 106, 108, 120, 132, 139,
    140, 242
environmental performance xiii
excellence factors 153, 154
external stakeholders 6

F
facilitator 52
facility auditing and corrective ac-
    tions 33
force field analysis 32
functional teams 24, 63

G
goal/objective 139
goals 12, 21, 27, 50, 221, 223

H
human factors 2

I
indicators 56

inspections 38
intensity indicator 54
internal audits 24, 25, 33, 38, 71,
    142, 149, 154, 164, 165
ISO 50001 Energy Management
    System (EnMS) 1, 45, 71, 73,
    75, 100, 139, 143, 144, 150,
    151, 179, 190, 236
  using 242
IT power management 232
  program 222

K
key performance indicators 23
key results areas 21, 23, 72

L
leadership 9, 12, 153, 241
  assessment 10, 11
legal 71
legal and other requirement 141
load shedding 72
low hanging fruit 154, 163, 170

M
management/employee energy
    brainstorming sessions 139
management/executive reviews 26
management representative 26, 41,
    238
management reviews 25, 33, 37, 38,
    145, 149, 150, 154, 163, 165,
    178, 185, 223, 242
measurements 23, 142
measures 15
monitoring and measurement 33

N
non-conformances 176

O
objective 21, 33, 41
objective/goal 139
objectives and targets 26, 35, 38, 48,
    50, 51, 72, 85, 92, 145, 149,
    154, 160, 164, 170, 224, 241,
    243
  and action plans 91, 99
office paper projects 232
operational controls 141, 142, 145
organizational culture 14, 187, 235

P
payback 45
payback analysis 218
peak load 58
performance indicators 21, 76
performance measurement 21, 71
plan of action 11
policy 11
power usage effectiveness (PUE)
    216, 219
preventative maintenance 27, 50, 65
preventative or corrective actions
    27, 103
procurement plans 72, 180
procurement policy 159
project management 242

R
recommissioning 74, 99, 228
reduction goal 224
reliability 120
renewable energy 59, 228
renewal 57
  energy goal 69
roles and responsibilities 33, 165,
    228

S
self inspections 24, 33, 37, 142, 149,
        154, 164, 176, 242
    checklist 33
sheddable 58
significant electricity users (SEUs)
        173, 174
significant energy users 100
SMART 85
special energy xi
strategies 11, 12, 41, 69, 88, 228
strengths, weaknesses, opportuni-
        ties, and threats (SWOT) 72
    analysis 44
Superior Energy Performance™
        144

T
targets and their action plans 33
team leader 34
teamwork 24
training 33, 63, 72, 154
    plan 179

V
verification 24

W
walkthroughs 61, 62, 64, 72, 74,
        179, 224
    team 95